PERIODIC TABLE
OF THE **ELEMENTS**

Halogens
and
Noble Gases

PERIODIC TABLE
OF THE **ELEMENTS**

Halogens
and
Noble Gases

Monica Halka, Ph.D., and
Brian Nordstrom, Ed.D.

Facts On File
An imprint of Infobase Publishing

HALOGENS AND NOBLE GASES

Facts On File, Inc.
An imprint of Infobase Publishing
132 West 31st Street
New York NY 10001

Library of Congress Cataloging-in-Publication Data
Halka, Monica.
  Halogens and noble gases / Monica Halka and Brian Nordstrom.
      p. cm. — (Periodic table of the elements)
  Includes bibliographical references and index.
  ISBN 978-0-8160-7368-9
  1. Halogens. 2. Gases, Rare. 3. Periodic law. I. Nordstrom, Brian. II. Title.
  QD165.H37 2010
  546'.73—dc22   2009031088

Facts On File books are available at special discounts when purchased in bulk quantities for businesses, associations, institutions, or sales promotions. Please call our Special Sales Department in New York at (212) 967-8800 or (800) 322-8755.

You can find Facts On File on the World Wide Web at http://www.factsonfile.com

Excerpts included herewith have been reprinted by permission of the copyright holders; the author has made every effort to contact copyright holders. The publishers will be glad to rectify, in future editions, any errors or omissions brought to their notice.

Text design by Erik Lindstrom
Composition by Hermitage Publishing Services
Illustrations by Richard Garratt
Photo research by Tobi Zausner, Ph.D.
Cover printed by Bang Printing, Brainerd, Minn.
Book printed and bound by Bang Printing, Brainerd, Minn.
Date printed: July 2010
Printed in the United States of America

10 9 8 7 6 5 4 3 2 1

This book is printed on acid-free paper.

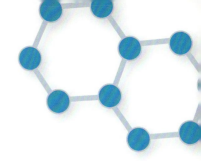

Contents

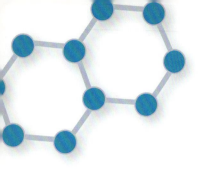

Preface

Speculations about the nature of matter date back to ancient Greek philosophers like Thales, who lived in the sixth century B.C.E., and Democritus, who lived in the fifth century B.C.E., and to whom we credit the first theory of *atoms*. It has taken two and a half millennia for natural philosophers and, more recently, for chemists and physicists to arrive at a modern understanding of the nature of *elements* and *compounds*. By the 19th century, chemists such as John Dalton of England had learned to define elements as pure substances that contain only one kind of atom. It took scientists like the British physicists Joseph John Thomson and Ernest Rutherford in the early years of the 20th century, however, to demonstrate what atoms are—entities composed of even smaller and more elementary particles called *protons, neutrons,* and *electrons.* These particles give atoms their properties and, in turn, give elements their physical and chemical properties.

After Dalton, there were several attempts throughout Western Europe to organize the known elements into a conceptual framework that would account for the similar properties that related groups of elements exhibit and for trends in properties that correlate with increases in atomic weights. The most successful *periodic table* of the elements was designed in 1869 by a Russian chemist, Dmitri Mendeleev. Mendeleev's method of organizing the elements into columns grouping elements with similar chemical and physical properties proved to be so practical that his table is still essentially the only one in use today.

While there are many excellent works written about the periodic table (which are listed in the section on further resources), recent scientific investigation has uncovered much that was previously unknown about nearly every element. The Periodic Table of the Elements, a six-volume set, is intended not only to explain how the elements were discovered and what their most prominent chemical and physical properties are, but also to inform the reader of new discoveries and uses in fields ranging from astrophysics to material science. Students, teachers, and the general public seldom have the opportunity to keep abreast of these new developments, as journal articles for the nonspecialist are hard to find. This work attempts to communicate new scientific findings simply and clearly, in language accessible to readers with little or no formal background in chemistry or physics. It should, however, also appeal to scientists who wish to update their understanding of the natural elements.

Each volume highlights a group of related elements as they appear in the periodic table. For each element, the set provides information regarding:

- the discovery and naming of the element, including its role in history, and some (though not all) of the important scientists involved;
- the basics of the element, including such properties as its atomic number, atomic mass, electronic configuration, melting and boiling temperatures, abundances (when known), and important isotopes;
- the chemistry of the element;
- new developments and dilemmas regarding current understanding; and
- past, present, and possible future uses of the element in science and technology.

Some topics, while important to many elements, do not apply to all. Though nearly all elements are known to have originated in stars or stellar explosions, little information is available for some. Some others that have been synthesized by scientists on Earth have not been observed in stellar spectra. If significant astrophysical nucleosynthesis research exists, it is presented as a separate section. The similar situation applies for geophysical research.

Special topic sections describe applications for two or more closely associated elements. Sidebars mainly refer to new developments of special interest. Further resources for the reader appear at the end of the book, with specific listings pertaining to each chapter, as well as a listing of some more general resources.

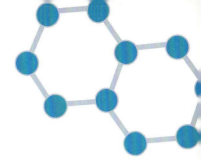

Acknowledgments

First and foremost, I thank my parents, who convinced me that I was capable of achieving any goal. In graduate school, my thesis adviser, Dr. Howard Bryant, influenced my way of thinking about science more than anyone else. Howard taught me that learning requires having the humility to doubt your understanding and that it is important for a physicist to be able to explain her work to anyone. I have always admired the ability to communicate scientific ideas to nonscientists and wish to express my appreciation for conversations with National Public Radio science correspondent Joe Palca, whose clarity of style I attempt to emulate in this work. I also thank my coworkers at Georgia Tech, Dr. Greg Nobles and Ms. Nicole Leonard, for their patience and humor as I struggled with deadlines.

—*Monica Halka*

In 1967, I entered the University of California at Berkeley. Several professors, including John Phillips, George Trilling, Robert Brown, Samuel Markowitz, and A. Starker Leopold, made significant and lasting impressions. I owe an especial debt of gratitude to Harold Johnston, who was my graduate research adviser in the field of atmospheric chemistry. I have known personally many of the scientists mentioned in the Periodic Table of the Elements set: For example, I studied under Neil Bartlett, Kenneth Street, Jr., and physics Nobel laureate Emilio Segrè. I especially cherish having known chemistry Nobel laureate Glenn

Seaborg. I also acknowledge my past and present colleagues at California State University; Northern Arizona University; and Embry-Riddle Aeronautical University, Prescott, Arizona, without whom my career in education would not have been as enjoyable.

—*Brian Nordstrom*

Both authors thank Jodie Rhodes and Frank Darmstadt for their encouragement, patience, and understanding.

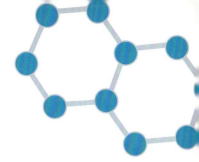

Introduction

Materials that are poor conductors of electricity are generally considered nonmetals. One important use of nonmetals, in fact, is their capability to insulate materials against the flow of electrical current. Earth's atmosphere is composed of nonmetallic elements, but lightning can break apart the chemical bonds and allow huge voltages to make their way to the ground. Water in its pure form is nonmetallic, though it almost always contains impurities called electrolytes that allow for an electric field.

While scientists categorize the chemical elements as nonmetals, metals, and metalloids—largely based on the elements' abilities to conduct electricity at normal temperatures and pressures—there are other distinctions taken into account when classifying the elements in the periodic table. The halogens, for example, are nonmetals, but have such special properties that they are given their own classification. The same is true for the noble gases. All the nonmetals, except hydrogen, appear on the right side of the periodic table (see the accompanying table on page xiv, "The Nonmetals Corner"). Hydrogen's place is usually shown at the upper left with the alkali metals, strictly because of its electron configuration, though it can be shown with the halogens and has been shifted in the following table for ease of grouping.

Halogens and Noble Gases presents the current scientific understanding of the physics, chemistry, geology, and biology of these two families of nonmetals, including how they are synthesized in the

THE NONMETALS CORNER

				H	He
B	C	N	O	F	Ne
(Al)	Si	P	S	Cl	Ar
(Ga)	Ge	As	Se	Br	Kr
(In)	(Sn)	Sb	Te	I	Xe
(Tl)	(Pb)	(Bi)	Po	At	Rn

Note: Halogens are in bold type. Noble gases are underlined. Metalloids are in italics. Post-transition metals are in parentheses.

universe, when and how they were discovered, and where they are found on Earth. The book also details how humans use halogens and noble gases and the resulting benefits and challenges to society, health, and the environment.

The first chapter is about the most chemically reactive element in the universe—fluorine. The extreme reactivity of elemental fluorine (as fluorine gas) makes it one of the most hazardous elements with which to work. At the same time, fluorine is a component of a variety of useful compounds. The compound hydrofluoric acid is used to etch glass, for example, and fluorine 18 is used in positron emission topography (PET), a nuclear medicine imaging technique.

Chapters, 2, 3, and 4 discuss chlorine, bromine, and iodine and astatine, respectively. Chlorine—in the form of sodium chloride, or *table salt*—is perhaps the most familiar halogen. The use of table salt to preserve and flavor foods dates to prehistoric times. Astatine, in all its forms, is radioactive, so that very little of it exists. On the other hand, the oceans are full of chlorine, bromine, and iodine, and large industries have grown up to extract these elements from seawater. As pure elements, the halogens are all toxic. As ions, however, chlorine and iodine are essential nutrients.

Chapter 5 examines the most chemically inert element in the universe—helium. Helium is also the second most abundant element in the universe, formed both by the fusion of hydrogen nuclei in the seething cauldrons of stellar cores and by the radioactive decay of uranium and other heavy metals. Helium may exhibit no chemistry, but its physics lends it numerous useful applications. Neither underground reserves nor the minuscule quantity of helium found in Earth's atmosphere can sustain humanity's current use of the gas. Like oil and natural gas, helium is a nonrenewable commodity that some experts predict the world will use up within 25 years.

Chapters 6, 7, and 8 investigate neon, argon, and krypton and xenon, respectively. Neon and argon are chemically inert, but lights made of neon and other noble gases illuminate the night sky. Argon, which is formed by the radioactive decay of potassium in Earth's crust, comprises 1 percent of the atmosphere. Krypton and xenon are found in much smaller quantities than either neon or argon, but they are capable of forming a number of chemical compounds.

Chapter 9 discusses radon, the heaviest noble gas and the only one that exists strictly in radioactive form. Radon is produced as a "great-granddaughter" of uranium; uranium decays into thorium, which decays into radium, which in turn decays into radon. Radon does not last very long, however, before it decays into polonium, since even radium's longest-lived isotope has a half-life of only 3.8 days. Radon is unique in the uranium-decay series in that it is the only gaseous element in the series; the rest are all solids.

Chapter 10 explains the fundamentals of chemistry and physics that explain the family properties of the halogens and noble gases. In addition, it presents possible future developments in halogen and noble gas science and its potential applications.

In spite of their adjacency in the periodic table, the properties of halogens and nonmetals are very different. The halogens are among the most chemically reactive elements in the periodic table and exhibit a diverse chemistry in terms of the large numbers of compounds they can form. On the other hand, noble gases are the least chemically reactive elements. In fact, before the 1960s, chemists referred to these elements as the *inert gases,* because it was believed that they exhibit no chemistry whatsoever. It was discovered, however, that krypton and xenon are capable of bonding with other elements, principally fluorine. (Even today, compounds have not been formed of helium, neon, or argon.)

As an important introductory tool, the reader should note the following properties of halogens and noble gases that show how they are similar but also how they differ:

1. The atoms of halogens and noble gases (as with nonmetals in general) tend to be smaller than those of metals. Several of the other properties of nonmetals result from their atomic sizes.
2. Nonmetals like halogens and noble gases exhibit very low electrical conductivities. The low, or nonexistent, electrical conductivity is the most important property that distinguishes nonmetals from metals.
3. Halogens have high electronegativities. This means that the atoms of halogens have a strong tendency to attract more electrons than might be expected. In contrast, noble gases exhibit almost no tendency to attract additional electrons.
4. Halogens have high electron affinities. This means that it is energetically very favorable to have their atoms gain additional electrons. In contrast, noble gases have negligible electron affinity.
5. Under normal conditions of temperature and pressure, most elements in their pure forms—including almost all metals (with the exception of mercury) and metalloids—exist as solids. In contrast, as their name implies, all of the noble gases are gases. There is a trend with the halogens that is exhibited as the column is descended: Fluorine and chlorine are gases, bromine is a liquid, and iodine and astatine are solids. The fact that so many of the elements in these two families exist as liquids or gases means that they generally have relatively low melting and boiling points under normal atmospheric conditions.
6. In their solid state, nonmetals like halogens and noble gases tend to be brittle. Therefore, they lack the malleability and ductility exhibited by metals.

The following is a list of the general chemical properties of halogens and noble gases, showing both how they are similar and how they are different:

1. Halogens are almost never found in nature as pure elements, whereas the noble gases exist in nature only as pure elements.

2. As pure elements, halogens exist only as diatomic molecules, for example, F_2, Cl_2, and Br_2. Noble gases are only monatomic species, for example, He, Ne, and Ar.

3. In aqueous solution, halogens form simple negative ions (called *halide* ions). These ions easily form ionic compounds with virtually all the metals. Noble gases do not form ions at all in aqueous solution.

4. With the exception of fluorine (which only forms the F^- ion), the halogens can form polyatomic, or complex, negative ions. Examples of polyatomic ions are ClO^-, ClO_4^-, and BrO_3^-. Again, noble gases do not form ions.

5. Halogens form covalent chemical bonds with other nonmetallic elements. Consequently, compounds of nonmetals often exist as small molecules. Halogens may also form covalent bonds with some of the metals or metalloids. In contrast, only krypton and xenon form such molecules. (Radon probably does, too, but few chemists study it because of its radioactivity.)

6. With the exception of fluorine, halogens can exist in both positive and negative *oxidation states*. This means, for example, that halogens tend to readily form compounds with both hydrogen and oxygen.

Halogens and Noble Gases provides the reader, whether student or scientist, with an up-to-date understanding regarding each of these families—where they came from, how they fit into our current technological society, and where they may lead us.

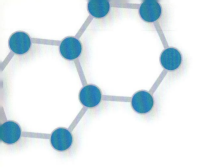

Overview: Chemistry and Physics Background

What *is* an element? To the ancient Greeks, everything on Earth was made from only four elements—earth, air, fire, and water. Celestial bodies—the Sun, moon, planets, and stars—were made of a fifth element: ether. Only gradually did the concept of an element become more specific.

An important observation about nature was that substances can change into other substances. For example, wood burns, producing heat, light, and smoke and leaving ash. Pure metals like gold, copper, silver, iron, and lead can be smelted from their ores. Grape juice can be fermented to make wine and barley fermented to make beer. Food can be cooked; food can also putrefy. The baking of clay converts it into bricks and pottery. These changes are all examples of chemical reactions. Alchemists' careful observations of many chemical reactions greatly helped them to clarify the differences between the most elementary substances ("elements") and combinations of elementary substances ("compounds" or "mixtures").

Elements came to be recognized as simple substances that cannot be decomposed into other even simpler substances by chemical reactions. Some of the elements that had been identified by the Middle Ages are easily recognized in the periodic table because they still have chemical symbols that come from their Latin names. These elements are listed in the table on page xix.

Russian chemist Dmitri Mendeleev created the periodic table of the elements in the late 1800s. *(Scala/ Art Resource)*

Modern atomic theory began with the work of the English chemist John Dalton in the first decade of the 19th century. As the concept of the atomic composition of matter developed, chemists began to define elements as simple substances that contain only one kind of atom. Because scientists in the 19th century lacked any experimental apparatus capable

ELEMENTS KNOWN TO ANCIENT PEOPLE

Iron: Fe ("ferrum")	Copper: Cu ("cuprum")
Silver: Ag ("argentum")	Gold: Au ("aurum")
Lead: Pb ("plumbum")	Tin: Sn ("stannum")
Antimony: Sb ("stibium")	Mercury: Hg ("hydrargyrum")
*Sodium: Na ("natrium")	*Potassium: K ("kalium")
Sulfur: S ("sulfur")	

Note: *Sodium and potassium were not isolated as pure elements until the early 1800s, but some of their salts were known to ancient people.

of probing the structure of atoms, the 19th-century model of the atom was rather simple. Atoms were thought of as small spheres of uniform density; atoms of different elements differed only in their masses. Despite the simplicity of this model of the atom, it was a great step forward in our understanding of the nature of matter. Elements could be defined as simple substances containing only one kind of atom. Compounds are simple substances that contain more than one kind of atom. Because atoms have definite masses, and only whole numbers of atoms can combine to make molecules, the different elements that make up compounds are found in definite proportions by mass. (For example, a molecule of water contains one oxygen atom and two hydrogen atoms, or a mass ratio of oxygen-to-hydrogen of about 8:1.) Since atoms are neither created nor destroyed during ordinary chemical reactions ("ordinary" meaning in contrast to "nuclear" reactions), what happens in chemical reactions is that atoms are rearranged into combinations that differ from the original reactants, but in doing so, the total mass is conserved. Mixtures are combinations of elements that are not in definite proportions. (In salt water, for example, the salt could be 3 percent by mass, or 5 percent by mass, or many other possibilities; regardless of the percentage of salt, it would still be called "salt water.") Chemical reactions are not required to separate the components of mixtures; the components of mixtures can be separated by physical processes such as distillation, evaporation, or precipitation. Examples of elements, compounds, and mixtures are listed in the following table.

EXAMPLES OF ELEMENTS, COMPOUNDS, AND MIXTURES

ELEMENTS	COMPOUNDS	MIXTURES
Hydrogen	Water	Salt water
Oxygen	Carbon dioxide	Air
Carbon	Propane	Natural gas
Sodium	Table salt	Salt and pepper
Iron	Hemoglobin	Blood
Silicon	Silicon dioxide	Sand

The definition of an element became more precise at the dawn of the 20th century with the discovery of the proton. We now know that an atom has a small center called the "nucleus." In the nucleus are one or more protons, positively charged particles, the number of which determine an atom's identity. The number of protons an atom has is referred to as its "atomic number." Hydrogen, the lightest element, has an atomic number of 1, which means each of its atoms contains a single proton. The next element, helium, has an atomic number of 2, which means each of its atoms contain two protons. Lithium has an atomic number of 3, so its atoms have three protons, and so forth, all the way through the periodic table. Atomic nuclei also contain neutrons, but atoms of the same element can have different numbers of neutrons; we call atoms of the same element with different number of neutrons "isotopes."

There are roughly 92 naturally occurring elements—hydrogen through uranium. Of those 92, two elements, technetium (element 43) and promethium (element 61), may once have occurred naturally on Earth, but the atoms that originally occurred on Earth have decayed away, and those two elements are now produced artificially in nuclear reactors. In fact, technetium is produced in significant quantities because of its daily use by hospitals in nuclear medicine. Some of the other first 92 elements—polonium, astatine, and francium, for example—are so radioactive that they exist in only tiny amounts. All of the elements with atomic numbers greater than 92—the so-called transuranium elements—are all produced artificially in nuclear reactors or particle accelerators. As of the writing of this book, the discoveries of the elements through number 118 (with the exception of number 117) have all been reported. The discoveries of elements with atomic numbers greater than 111 have not yet been confirmed, so those elements have not yet been named.

When the Russian chemist Dmitri Mendeleev (1834–1907) developed his version of the periodic table in 1869, he arranged the elements known at that time in order of *atomic mass* or *atomic weight* so that they fell into columns called *groups* or *families* consisting of elements with similar chemical and physical properties. By doing so, the rows exhibit periodic trends in properties going from left to right across the table, hence the reference to rows as *periods* and name "periodic table."

Mendeleev's table was not the first periodic table, nor was Mendeleev the first person to notice *triads* or other groupings of elements with similar properties. What made Mendeleev's table successful and the one we use today are two innovative features. In the 1860s, the concept of *atomic number* had not yet been developed, only the concept of atomic mass. Elements were always listed in order of their atomic masses, beginning with the lightest element, hydrogen, and ending with the heaviest element known at that time, uranium. Gallium and germanium, however, had not yet been discovered. Therefore, if one were listing the known elements in order of atomic mass, arsenic would follow zinc, but that would place arsenic between aluminum and indium. That does not make sense because arsenic's properties are much more like those of phosphorus and antimony, not like those of aluminum and indium.

To place arsenic in its "proper" position, Mendeleev's first innovation was to leave two blank spaces in the table after zinc. He called the first element eka-aluminum and the second element eka-silicon,

Mendeleev's Periodic Table (1871)

Group / Period	I	II	III	IV	V	VI	VII	VIII
1	H=1							
2	Li=7	Be=9.4	B=11	C=12	N=14	O=16	F=19	
3	Na=23	Mg=24	Al=27.3	Si=28	P=31	S=32	Cl=35.5	
4	K=39	Ca=40	?=44	Ti=48	V=51	Cr=52	Mn=55	Fe=56, Co=59 Ni=59
5	Cu=63	Zn=65	?=68	?=72	As=75	Se=78	Br=80	
6	Rb=85	Sr=87	?Yt=88	Zr=90	Nb=94	Mo=96	?=100	Ru=104, Rh=104 Pd=106
7	Ag=108	Cd=112	In=113	Sn=118	Sb=122	Te=125	J=127	
8	Cs=133	Ba=137	?Di=138	?Ce=140				
9								
10			?Er=178	?La=180	Ta=182	W=184		Os=195, Ir=197 Pt=198
11	Au=199	Hg=200	Tl=204	Pb=207	Bi=208			
12				Th=231		U=240		

© Infobase Publishing

Dmitri Mendeleev's 1871 periodic table. The elements listed are the ones that were known at that time, arranged in order of increasing relative atomic mass. Mendeleev predicted the existence of elements with masses of 44, 68, and 72. His predictions were later shown to have been correct.

which he said corresponded to elements that had not yet been discovered but whose properties would resemble the properties of aluminum and silicon, respectively. Not only did Mendeleev predict the elements' existence, he also estimated what their physical and chemical properties should be in analogy to the elements near them. Shortly afterward, these two elements were discovered and their properties were found to be very close to what Mendeleev had predicted. Eka-aluminum was called *gallium* and eka-silicon was called *germanium*. These discoveries validated the predictive power of Mendeleev's arrangement of the elements and demonstrated that Mendeleev's periodic table could be a predictive tool, not just a compendium of information that people already knew.

The second innovation Mendeleev made involved the relative placement of tellurium and iodine. If the elements are listed in strict order of their atomic masses, then iodine should be placed before tellurium, since iodine is lighter. That would place iodine in a group with sulfur and selenium and tellurium in a group with chlorine and bromine, an arrangement that does not work for either iodine or tellurium. Therefore, Mendeleev rather boldly reversed the order of tellurium and iodine so that tellurium falls below selenium and iodine falls below bromine. More than 40 years later, after Mendeleev's death, the concept of atomic number was introduced, and it was recognized that elements should be listed in order of atomic number, not atomic mass. Mendeleev's ordering was thus vindicated, since tellurium's atomic number is one less than iodine's atomic number. Before he died, Mendeleev was considered for the Nobel Prize, but did not receive sufficient votes to receive the award despite the importance of his insights.

THE PERIODIC TABLE TODAY

All of the elements in the first 12 groups of the periodic table are referred to as *metals*. The first two groups of elements on the left-hand side of the table are the *alkali metals* and the *alkaline earth metals*. All of the alkali metals are extremely similar to each other in their chemical and physical properties, as, in turn, are all of the alkaline earths to each other. The 10 groups of elements in the middle of the periodic table are *transition metals*. The similarities in these groups are not as strong as those in the first two groups, but still satisfy the general trend of similar chemical and physical properties. The transition metals in the last row are not found in nature but have been synthesized artificially. The metals that follow the transition metals are called post-transition metals.

The so-called *rare earth elements*, which are all metals, usually are displayed in a separate block of their own located below the rest of the periodic table. The elements in the first row of rare earths are called *lanthanides* because their properties are extremely similar to the properties of lanthanum. The elements in the second row of rare earths are called *actinides* because their properties are extremely similar to the properties of actinium. The actinides following uranium are called *transuranium elements* and are not found in nature but have been produced artificially.

The far right-hand six groups of the periodic table—the remaining *main group elements*—differ from the first 12 groups in that more than one kind of element is found in them; in this part of the table we find metals, all of the *metalloids* (or *semimetals*), and all of the *nonmetals.* Not counting the artificially synthesized elements in these groups (elements having atomic numbers of 113 and above and that have not yet been named), these six groups contain 7 metals, 8 metalloids, and 16 nonmetals. Except for the last group—the *noble gases*—each individual group has more than just one kind of element. In fact, sometimes nonmetals, metalloids, and metals are all found in the same column, as are the cases with group IVB (C, Si, Ge, Sn, and Pb) and also with group VB (N, P, As, Sb, and Bi). Although similarities in chemical and physical properties are present within a column, the differences are often more striking than the similarities. In some cases, elements in the same column do have very similar chemistry. Triads of such elements include three of the *halogens* in group VIIB—chlorine, bromine, and iodine; and three group VIB elements—sulfur, selenium, and tellurium.

ELEMENTS ARE MADE OF ATOMS

An atom is the fundamental unit of matter. In ordinary chemical reactions, atoms cannot be created or destroyed. Atoms contain smaller *subatomic* particles: protons, neutrons, and electrons. Protons and neutrons are located in the *nucleus,* or center, of the atom and are referred to as *nucleons.* Electrons are located outside the nucleus. Protons and neutrons are comparable in mass and significantly more massive than electrons. Protons carry positive electrical charge. Electrons carry negative charge. Neutrons are electrically neutral.

The identity of an element is determined by the number of protons found in the nucleus of an atom of the element. The number of protons is called an element's atomic number, and is designated by the letter Z. For hydrogen, $Z = 1$, and for helium, $Z = 2$. The heaviest naturally occurring element is uranium, with $Z = 92$. The value of Z is 118 for the heaviest element that has been synthesized artificially.

Atoms of the same element can have varying numbers of neutrons. The number of neutrons is designated by the letter N. Atoms of the same element that have different numbers of neutrons are called *isotopes* of that element. The term *isotope* means that the atoms occupy the same place in the periodic table. The sum of an atom's protons and neutrons is called the atom's *mass number*. Mass numbers are dimensionless whole numbers designated by the letter A and should not be confused with an atom's *mass,* which is a decimal number expressed in units such as grams. Most elements on Earth have more than one isotope. The average mass number of an element's isotopes is called the element's atomic mass or atomic weight.

The standard notation for designating an atom's atomic and mass numbers is to show the atomic number as a subscript and the mass number as a superscript to the left of the letter representing the element. For example, the two naturally occurring isotopes of hydrogen are written $^{1}_{1}H$ and $^{2}_{1}H$.

For atoms to be electrically neutral, the number of electrons must equal the number of protons. It is possible, however, for an atom to gain or lose electrons, forming *ions.* Metals tend to lose one or more electrons to form positively charged ions (called *cations*); nonmetals are more likely to gain one or more electrons to form negatively charged ions (called *anions*). Ionic charges are designated with superscripts. For example, a calcium ion is written as Ca^{2+}; a chloride ion is written as Cl^{-}.

THE PATTERN OF ELECTRONS IN AN ATOM

During the 19th century, when Mendeleev was developing his periodic table, the only property that was known to distinguish an atom of one element from an atom of another element was relative mass. Knowledge of atomic mass, however, did not suggest any relationship between an element's mass and its properties. It took several discoveries—among them that of the electron in 1897 by the British physicist John Joseph ("J. J.") Thomson, *quanta* in 1900 by the German physicist Max Planck, the wave nature of matter in 1923 by the French physicist Louis de Broglie, and the mathematical formulation of the quantum mechanical model of the atom in 1926 by the German physicists Werner Heisenberg and Erwin Schrödinger (all of whom collectively illustrate the international nature of science)—to elucidate the relationship between the structures of atoms and the properties of elements.

The number of protons in the nucleus of an atom defines the identity of that element. Since the number of electrons in a neutral atom is equal to the number of protons, an element's atomic number also reveals how many electrons are in that element's atoms. The electrons occupy regions of space that chemists and physicists call *shells.* The shells are further divided into regions of space called *subshells.* Subshells are related to angular momentum, which designates the shape of the electron orbit space around the nucleus. Shells are numbered 1, 2, 3, 4, and so forth (in theory out to infinity). In addition, shells may be designated by letters: The first shell is the K-shell, the second shell the L-shell, the third the M-shell, and so forth. Subshells have letter designations, s, p, d, and f being the most common. The *n*th shell has *n* possible subshells. Therefore, the first shell has only an s subshell, designated 1s; the second shell has both s and p subshells (2s and 2p); the third shell 3s, 3p, and 3d; and the fourth shell 4s, 4p, 4d, and 4f. (This pattern continues for higher-numbered shells, but this is enough for now.)

An s subshell is spherically symmetric and can hold a maximum of 2 electrons. A p subshell is dumbbell-shaped and holds 6 electrons, a d subshell 10 electrons, and an f subshell 14 electrons, with increasingly complicated shapes.

As the number of electrons in an atom increases, so does the number of shells occupied by electrons. In addition, because electrons are all negatively charged and tend to repel each other *electrostatically,* as the number of the shell increases, the size of the shell increases, which means that electrons in higher-numbered shells are located, on the average, farther from the nucleus. Inner shells tend to be fully occupied with the maximum number of electrons they can hold. The electrons in

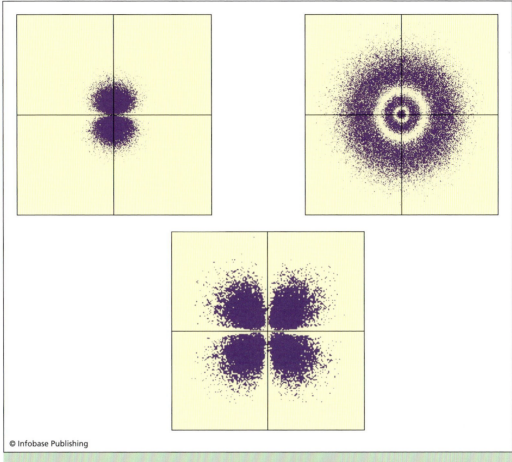

© Infobase Publishing

Some hydrogen wavefunction distributions for electrons in various excited states

the outermost shell, which is likely to be only partially occupied, will determine that atom's properties.

Physicists and chemists use *electronic configurations* to designate which subshells in an atom are occupied by electrons as well as how many electrons are in each subshell. For example, nitrogen is element number 7, so it has seven electrons. Nitrogen's electronic configuration is $1s^2 2s^2 2p^3$; a superscript designates the number of electrons that occupy a subshell. The first shell is fully occupied with its maximum of two electrons. The second shell can hold a maximum of eight electrons, but it is only partially occupied with just five electrons—two in the 2s sub-shell and three in the 2p. Those five outer electrons determine nitrogen's properties. For a heavy element like tin (Sn), electronic configurations can be quite complex. Tin's configuration is $1s^2 2s^2 2p^6 3s^2 3p^6 4s^2 3d^{10} 4p^6 5s^2 4d^{10} 5p^2$ but is more commonly written in the shorthand notation [Kr] $5s^2 4d^{10} 5p^2$, where [Kr] represents the electron configuration pattern for the noble gas krypton. (The pattern continues in this way for shells with higher numbers.) The important thing to notice about tin's configuration is that all of the shells except the last one are fully occupied. The fifth shell can hold 32 electrons, but in tin there are only four electrons in the fifth shell. The outer electrons determine an element's properties. The table on page xxix illustrates the electronic configurations for nitrogen and tin.

ATOMS ARE HELD TOGETHER WITH CHEMICAL BONDS

Fundamentally, a chemical bond involves either the sharing of two electrons or the transfer of one or more electrons to form ions. Two atoms of nonmetals tend to share pairs of electrons in what is called a *covalent bond*. By sharing electrons, the atoms remain more or less electrically neutral. However, when an atom of a metal approaches an atom of a nonmetal, the more likely event is the transfer of one or more electrons from the metal atom to the nonmetal atom. The metal atom becomes a positively charged ion and the nonmetal atom becomes a negatively charged ion. The attraction between opposite charges provides the force that holds the atoms together in what is called an *ionic bond*. Many chemical bonds are also intermediate in nature between covalent and ionic bonds and have characteristics of both types of bonds.

IN CHEMICAL REACTIONS, ATOMS REARRANGE TO FORM NEW COMPOUNDS

When a substance undergoes a *physical change,* the substance's name does not change. What may change is its temperature, its length, its *physical state* (whether it is a solid, liquid, or gas), or some other characteristic, but it is still the same substance. On the other hand, when a substance undergoes a *chemical change,* its name changes; it is a different substance. For example, water can decompose into hydrogen gas and oxygen gas, each of which has substantially different properties from water, even though water is composed of hydrogen and oxygen atoms.

ELECTRONIC CONFIGURATIONS FOR NITROGEN AND TIN

ELECTRONIC CONFIGURATION OF NITROGEN (7 ELECTRONS)

Energy Level	Shell	Subshell	Number of Electrons
1	K	s	2
2	L	s	2
		p	3
			7

ELECTRONIC CONFIGURATION OF TIN (50 ELECTRONS)

Energy Level	Shell	Subshell	Number of Electrons
1	K	s	2
2	L	s	2
		p	6
3	M	s	2
		p	6
		d	10
4	N	s	2
		p	6
		d	10
5	O	s	2
		p	2
			50

In chemical reactions, the atoms themselves are not changed. Elements (like hydrogen and oxygen) may combine to form compounds (like water), or compounds can be decomposed into their elements. The atoms in compounds can be rearranged to form new compounds whose names and properties are different from the original compounds. Chemical reactions are indicated by writing chemical equations such as the equation showing the decomposition of water into hydrogen and oxygen: $2 H_2O$ (*l*) $\rightarrow 2 H_2$ (g) + O_2 (g). The arrow indicates the direction in which the reaction proceeds. The reaction begins with the *reactants* on the left and ends with the *products* on the right. We sometimes designate the physical state of a reactant or product in parentheses—*s* for solid, *l* for liquid, *g* for gas, and *aq* for *aqueous* solution (in other words, a solution in which water is the solvent).

IN NUCLEAR REACTIONS THE NUCLEI OF ATOMS CHANGE

In ordinary chemical reactions, chemical bonds in the reactant species are broken, the atoms rearrange, and new chemical bonds are formed in the product species. These changes only affect an atom's electrons; there is no change to the nucleus. Hence there is no change in an element's identity. On the other hand, nuclear reactions refer to changes in an atom's nucleus (whether or not there are electrons attached). In most nuclear reactions, the number of protons in the nucleus changes, which means that elements are changed, or transmuted, into different elements. There are several ways in which *transmutation* can occur. Some transmutations occur naturally, while others only occur artificially in nuclear reactors or particle accelerators.

The most familiar form of transmutation is *radioactive decay,* a natural process in which a nucleus emits a small particle or *photon* of light. Three common modes of decay are labeled *alpha, beta,* and *gamma* (the first three letters of the Greek alphabet). Alpha decay occurs among elements at the heavy end of the periodic table, basically elements heavier than lead. An alpha particle is a nucleus of helium 4 and is symbolized as $_2^4$He or α. An example of alpha decay occurs when uranium 238 emits an alpha particle and is changed into thorium 234 as in the following reaction: $_{92}^{238}U \rightarrow {}_2^4He + {}_{90}^{234}Th$. Notice that the parent isotope, U-238, has 92 protons, while the daughter isotope, Th-234, has only 90 protons.

The decrease in the number of protons means a change in the identity of the element. The mass number also decreases.

Any element in the periodic table can undergo beta decay. A beta particle is an electron, commonly symbolized as β^- or e^-. An example of beta decay is the conversion of cobalt 60 into nickel 60 by the following reaction: $^{60}_{27}Co \rightarrow {}^{60}_{28}Ni + e^-$. The atomic number of the daughter isotope is one greater than that of the parent isotope, which maintains charge balance. The mass number, however, does not change.

In gamma decay, photons of light (symbolized by γ) are emitted. Gamma radiation is a high-energy form of light. Light carries neither mass nor charge, so the isotope undergoing decay does not change identity; it only changes its energy state.

Elements also are transmuted into other elements by nuclear *fission* and *fusion*. Fission is the breakup of very large nuclei (at least as heavy as uranium) into smaller nuclei, as in the fission of U-236 in the following reaction: $^{236}_{92}U \rightarrow {}^{94}_{36}Kr + {}^{139}_{56}Ba + 3n$, where n is the symbol for a neutron (charge = 0, mass number = +1). In fusion, nuclei combine to form larger nuclei, as in the fusion of hydrogen isotopes to make helium. Energy may also be released during both fission and fusion. These events may occur naturally—fusion is the process that powers the Sun and all other stars—or they may be made to occur artificially.

Elements can be transmuted artificially by bombarding heavy target nuclei with lighter projectile nuclei in reactors or accelerators. The transuranium elements have been produced that way. Curium, for example, can be made by bombarding plutonium with alpha particles. Because the projectile and target nuclei both carry positive charges, projectiles must be accelerated to velocities close to the speed of light to overcome the force of repulsion between them. The production of successively heavier nuclei requires more and more energy. Usually, only a few atoms at a time are produced.

ELEMENTS OCCUR WITH DIFFERENT RELATIVE ABUNDANCES

Hydrogen overwhelmingly is the most abundant element in the universe. Stars are composed mostly of hydrogen, followed by helium and only very small amounts of any other element. Relative abundances of elements can be expressed in parts per million, either by mass or by numbers of atoms.

On Earth, elements may be found in the lithosphere (the rocky, solid part of Earth), the hydrosphere (the aqueous, or watery, part of Earth), or the atmosphere. Elements such as the noble gases, the rare earths, and commercially valuable metals like silver and gold occur in only trace quantities. Others, like oxygen, silicon, aluminum, iron, calcium, sodium, hydrogen, sulfur, and carbon are abundant.

HOW NATURALLY OCCURRING ELEMENTS HAVE BEEN DISCOVERED

For the elements that occur on Earth, methods of discovery have been varied. Some elements—like copper, silver, gold, tin, and lead—have been known and used since ancient or even prehistoric times. The origins of their early metallurgy are unknown. Some elements, like phosphorus, were discovered during the Middle Ages by alchemists who recognized that some mineral had an unknown composition. Sometimes, as in the case of oxygen, the discovery was by accident. In other instances—as in the discoveries of the alkali metals, alkaline earths, and lanthanides—chemists had a fairly good idea of what they were looking for and were able to isolate and identify the elements quite deliberately.

To establish that a new element has been discovered, a sample of the element must be isolated in pure form and subjected to various chemical and physical tests. If the tests indicate properties unknown in any other element, it is a reasonable conclusion that a new element has been discovered. Sometimes there are hazards associated with isolating a substance whose properties are unknown. The new element could be toxic, or so reactive that it can explode, or extremely radioactive. During the course of history, attempts to isolate new elements or compounds have resulted in more than just a few deaths.

HOW NEW ELEMENTS ARE MADE

Some elements do not occur naturally, but can be synthesized. They can be produced in nuclear reactors, from collisions in particle accelerators, or can be part of the *fallout* from nuclear explosions. One of the elements most commonly made in nuclear reactors is technetium. Relatively large quantities are made every day for applications in nuclear medicine. Sometimes, the initial product made in an accelerator is a heavy element whose atoms have very short *half-lives* and undergo radioactive decay. When the atoms decay, atoms of elements lighter than the parent atoms are produced. By identifying the daughter atoms, scientists can work backward and correctly identify the parent atoms from which they came.

The major difficulty with synthesizing heavy elements is the number of protons in their nuclei ($Z > 92$). The large amount of positive charge makes the nuclei unstable so that they tend to disintegrate either by radioactive decay or *spontaneous fission*. Therefore, with the exception of a few transuranium elements like plutonium (Pu) and americium (Am), most artificial elements are made only a few atoms at a time and so far have no practical or commercial uses.

THE HALOGENS AND NOBLE GASES SECTION OF THE PERIODIC TABLE

The book has been separated into the following two sections:

1. the halogens, and
2. the noble gases.

While both groups appear on the far right side of the periodic table, their chemical properties are very different, with the most notable characteristic of the noble gases being their essential nonreactivity.

Element		
K		M.P.°
L	**E**$_Z$	B.P.°
M		C.P.°
N		
O		
P	Oxidation states	
Q	Atomic weight	
	Abundance%	

Information box key. E represents the element's letter notation (for example, H = hydrogen), with the Z subscript indicating proton number. Orbital shell notations appear in the column on the left. For elements that are not naturally abundant, the mass number of the longest-lived isotope is given in brackets. The abundances (atomic %) are based on meteorite and solar wind data. The melting point (M.P.), boiling point (B.P.), and critical point (C.P.) temperatures are expressed in Celsius. Sublimation and critical point temperatures are indicated by *s* and *t*.

PART

I

The Halogens

INTRODUCTION TO THE HALOGENS

Nonmetals are distributed among five groups of elements in the periodic table—groups IVB, VB, VIB, VIIB, and VIII. The nonmetals in groups IVB, VB, and VIB are covered elsewhere in *Nonmetals,* another volume in this multivolume set. Those elements display a wide range of chemical and physical properties such that group trends are less apparent. The elements in this volume—group VIIB, the *halogens,* and group VIII, the *noble gases*—are much more similar to other elements in their same group, strongly exhibiting common group properties. Halogens are never found as pure elements, but when they are isolated, the halogens are all *diatomic* gases, which means that they consist of molecules that have two atoms in them. They are also powerful *oxidizing agents,* which means that they are very chemically reactive and tend to attack metals and other neutral elements. The name "halogen" is derived from Greek words related to the ability of these elements to form salts.

The following five elements are halogens:

1. fluorine (F),
2. chlorine (Cl),
3. bromine (Br),
4. iodine (I), and
5. astatine (At).

(Presumably ununseptium, element number 117, will be a halogen because it will be in the same column as the other halogens; however, element 117 has not yet been "discovered" as of the writing of this book.) Astatine occurs naturally on Earth in only minute quantities. All of its isotopes have very short half-lives; even when samples of astatine are produced artificially, they decay before very much astatine has had time to accumulate. On the other hand, appreciable quantities of the other halogens do occur naturally, but never as the free elements. In nature, fluorine and chlorine are the two most abundant halogens, but they exist almost exclusively as the *halide ions,* fluoride and chloride. Similarly, bromine and iodine are most likely to be found in the form of bromide and iodide ions, or else in organic compounds. Fluorine occurs mainly in the minerals fluorite (CaF_2) and cryolite (Na_3AlF_6). Because almost all halide compounds are appreciably soluble in water, the principal source of the other halogens—chlorine, bromine, and iodine—is seawater or marine organisms like kelp. *Sea salt* is principally composed of salts containing ions such as sodium (Na^+), potassium (K^+), chloride (Cl^-), bromide (Br^-), calcium (Ca^{2+}), magnesium (Mg^{2+}), sulfate (SO_4^{2-}), and bicarbonate (HCO_3^-).

The chemistry of fluorine, chlorine, bromine, and iodine is very similar. The differences that do exist are due primarily to the differences in sizes of the halogen atoms. Fluorine atoms are comparatively very small, and atomic size increases upon descending the column, with iodine atoms being the largest. Fluorine's much smaller size makes fluorine's chemistry less like the other three elements, which are very similar to each other. For example, HF is a weak acid, whereas HCl, HBr, and HI are strong *acids.* A fluorine atom can form only one chemical bond to other atoms, whereas chlorine, bromine, and iodine atoms are capable of forming several bonds to other atoms. The only ion fluorine can form is fluoride, whereas the other halogens can form oxyanions such as hypochlorite (ClO^-), perchlorate (ClO_4^-), bromate (BrO_4^-), and iodate (IO_4^-).

When they are present as free elements, all of the halogens are in the form of diatomic molecules. Fluorine (F_2) is extremely difficult to form from fluoride compounds. When it is made, it is in the gaseous state and is so reactive that it is extremely hazardous to handle. In fact, fluorine is the most reactive element in the periodic table. On the other hand, Cl_2, Br_2, and I_2 are relatively simple to form, although they are still reactive enough that they must be handled with caution. Under normal conditions, F_2 and Cl_2 are gaseous species, Br_2 is a *volatile* liquid, and I_2 is a volatile solid (at one atmosphere pressure and room temperature, solid iodine sublimes). Fluorine gas is a pale yellow color, chlorine gas is greenish-yellow, liquid bromine is a deep red color, and solid iodine (and its vapor) is violet.

The *oxidation state* of an element is a description of the chemical bonding of that element to other elements in a compound. (In simple cases, an ion's oxidation state is the same as its charge. In more complex cases, an element's oxidation state reflects how many covalent bonds it has formed to atoms of other elements.) The halogens are known principally as oxidizing agents. An oxidizing agent is a chemical substance that pulls electrons away from other elements, which in the case of the halogens is why they so readily form negative ions. (An element's relative tendency to pull electrons toward itself is called the element's *electronegativity.* Fluorine is the most electronegative element in the periodic table.) In contrast, a *reducing agent* is a chemical substance that donates electrons to atoms of other elements. An oxidizing agent itself undergoes *reduction,* and a reducing agent itself undergoes oxidation. The reaction between an oxidizing agent and a reducing agent is called an *oxidation-reduction reaction.*

The chemistry of all the halogens is dominated by the fact that they are among the most chemically reactive elements in the periodic table. In particular, the strong tendency of neutral halogen atoms to gain

electrons causes them to form ions with a charge of -1. If halogen molecules are to gain electrons, that must mean that atoms of other elements
are losing electrons. Because neutral halogen molecules have such a
strong tendency to gain electrons and be reduced, the halogens themselves are oxidizing agents. The strengths as oxidizing agents decrease
upon descending the column from fluorine to chlorine to bromine to
iodine, but each of the halogens is still considered to be a strong oxidizing agent. The most familiar example is the use of chlorine to disinfect
municipal water supplies.

Halogen oxyanions also are oxidizing agents. Just as Cl_2 is the principal substance used to disinfect municipal drinking water supplies,
the hypochlorite ion (ClO^-) is the oxidizing agent in household bleach.
Sometimes, chlorine dioxide (ClO_2) is used to disinfect water. In theory, all the halogens could be used to disinfect cuts and wounds. However, fluorine, chlorine, and bromine are too dangerous to apply to the
skin. Iodine is safer, so compounds containing iodine are used instead
(although it still tends to sting an open cut). The common disinfectant,
betadine, releases iodine in a relatively safe form, although people with
allergic sensitivities to iodine may need to avoid Betadine and other
sources of iodine. For example, shellfish tend to be high in iodine; a
person who eats shellfish and is highly allergic to iodine could go into
anaphylactic shock.

Combining halogens with most metals will oxidize the metal atoms
to higher oxidation states, thus forming compounds called metal halides.
In the meantime, the halogens are reduced to lower oxidation states.
When a metal atom has been oxidized to the +1 or +2 states, the metal
most likely has become an ion, and the bonding is ionic. If the metal has
been oxidized to a higher oxidation state, the bonding is more likely to
be covalent. Examples of ionic compounds are sodium chloride (NaCl),
calcium chloride ($CaCl_2$), lithium fluoride (LiF), and potassium iodide
(KI). Examples of covalent compounds are $FeCl_3$, $TiCl_4$, and $SnCl_4$.

Halogens also form *interhalogen compounds* (or halogen halides),
such as ClF, BrF_3, BrCl, IF_5, and IF_7. In general, these compounds are
powerful oxidizing and *halogenating agents*. Halogen fluorides readily
attack metals, often oxidizing the metal atoms to unusually high oxidation states. Examples include the formation of AgF_2 and CoF_3, which

are unusual because normally silver forms AgF and cobalt forms CoF_2. Halogen halides are very reactive with water, sometimes explosively. Some of the halogen halides can conduct electricity in the liquid state, although their electrical conductivities are typically much smaller than similar molten salts.

UNDERSTANDING PATTERNS AND PROPERTIES IN THE HALOGENS AND NOBLE GASES

The chemical and physical properties of elements are determined by their electronic configurations. Certain electronic configurations are especially stable with respect to their energies. This is the case with the noble gases, which, in general, have filled p subshells. With the exception of helium, which has an electronic configuration of $1s^2$ (which itself is a stable configuration), each noble gas element has a valence electron configuration of ns^2np^6, where n = 2 for neon, 3 for argon, 4 for krypton, and so forth. (This configuration for noble gases beginning with neon gives these elements eight valence electrons, which are referred to as a stable "octet" of electrons.)

The removal of one or more electrons from an atom of a noble gas requires an input of energy. As a general rule, that would be energetically unfavorable. (Adding an additional electron would also be unfavorable.) Therefore, when the noble gases were first discovered and thought not to form chemical compounds at all, they were called *inert* gases, the word "inert" meaning that a chemical substance is largely devoid of reactive properties. As the quantum mechanical model of the atom developed during the 1920s, chemists came to realize that the tendencies of the alkali metals and alkaline earths to form +1 and +2 ions, respectively, or the halogens to form -1 ions were driven by energetics. For example, ions such as Na^+, Mg^{2+}, and F^- would all be isoelectronic with the noble gas neon and would, therefore, share the energetic stability of a neon atom. Becoming "like a noble gas" was seen as a driving force for both ionic and covalent bonding. Chemists call this tendency the octet rule, since it results in atoms of these elements acquiring the same stable octets of valence electrons shared by the noble gases.

What distinguishes the properties of one noble gas element from another is the relative sizes of the atoms. As a general rule in the periodic table, upon descending a column of the table, the atoms in that family become relatively larger in size. This is because the valence electrons of elements toward the bottom of a column are in higher energy levels, or shells, and thus are farther from the nuclei than the valence electrons of elements toward the top of that column. Helium, in fact, has the smallest atoms of any element. Descending the column, the sizes of atoms increase in going from helium to neon to argon. However, all three of those noble gas elements only have electrons in s and p subshells. These electrons are quite close to the nuclei and are held very tightly to those atoms, both because of the small sizes of the atoms and because of the energetic favorability of having assumed an electronic configuration marked by a stable octet of eight valence electrons.

One consequence of the increase in the relative sizes of the atoms is an increase in melting points and boiling points. As shown in the graph below, the trend for the noble gases is fairly smooth, ranging from a boiling point of helium just above absolute zero to a boiling point of radon at temperatures that might be encountered in polar regions. The halogens exhibit a similarly smooth trend in increasing boiling points. In their case, however, the boiling points are respectively higher than

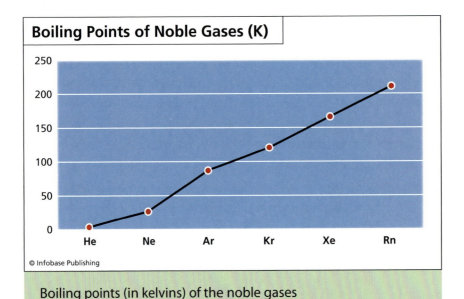

Boiling points (in kelvins) of the noble gases

those of the noble gases. Although F_2 and Cl_2 boil below room temperature and, therefore, are gases under normal atmospheric conditions, Br_2 and I_2 boil well above room temperature. The result is that Br_2 is a liquid under normal conditions, and I_2 is a solid.

Another consequence of the increase in the relative sizes of atoms is a change in the chemical reactivities of the elements in these two columns. Because krypton also has a subshell of d electrons—and xenon has subshells of both d and f electrons—their atoms have electrons that can participate in chemical bonding that do not affect the stability of the p subshell octets. Thus, these noble gas elements do form chemical compounds. Xenon has the most known examples because xenon has electrons at energy levels similar to some of the electrons in the heavier transition metals. In fact, it was this similarity that prompted Neil Bartlett in 1962 to attempt a reaction between xenon gas and fluorine gas, a reaction that proved to be successful. (Radon presumably should form similar to xenon's compounds, but much less research has been performed on radon's chemistry because of radon's radioactivity.)

In the case of the halogens, each element has an electronic configuration that is one electron short of the stable octet possessed by the noble gas just to the right of it in the table. (Each halogen has seven valence electrons with the configuration ns^2np^5.) Fluorine is the most reactive element in the family because the very small size of fluorine atoms causes a fluorine nucleus to exert an extremely strong force of attraction on an electron of a nearby atom. (This strong force of attraction is the reason fluorine is the most electronegative element in the periodic table.) The result is that a fluorine atom in contact with atoms of almost any other element will react so as to pull an electron either completely away, or at least partially away, from the other atom, resulting in either a fluorine ion or in a polar bond between the atoms. Doing so gives fluorine an effective electron configuration of $1s^22s^22p^6$, which is the same stable configuration that a neon atom has.

On the one hand, the chemical reactivities of the diatomic halogen molecules themselves decrease descending the column (due to the increase in the sizes of the molecules). On the other hand, the variety of chemical compounds that the halogen elements can form increases descending the column. Fluorine can only form compounds in which fluorine is in the -1 oxidation state. In the cases of the other elements, however, oxidation states of +1, +3, +5, +7, and even +4 are possible. These higher oxidation states are possible because the valence electrons of chlorine, bromine, and iodine can move into unoccupied d subshells and thus become available for a greater number of chemical bonds to atoms of other elements. The d subshell is too high in energy for the valence electrons in fluorine, so that a fluorine atom is limited to just one bond to other atoms.

This same trend in increasing sizes of atoms is true for the other families of elements in the periodic table. A corresponding trend is in decreasing sizes of atoms proceeding from left to right across a row (period) of the table. As a result, the largest atoms are found on the lower left-hand side of the table, where the large sizes are responsible for the chemical and physical properties (like high electrical conductivity) that we associate with elements that are metals. The smallest atoms are found on the upper right-hand side of the table, where the small sizes are associated with the properties of nonmetals (such as the property of being electrical insulators). Elements toward the interior of the table often exhibit the physical properties we associate with metals and the chemical properties we associate with nonmetals.

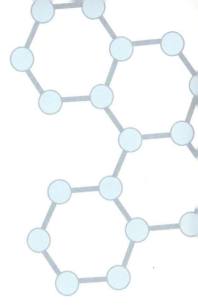

1

Fluorine: Corrosive, Toxic, and Remarkable

Fluorine is element number 9. It is a fairly common element—13th in order of abundance. In Earth's crust, it is found mostly in minerals that contain the fluoride ion (F^-). The fluoride ion, however, is extremely difficult to reduce safely to neutral fluorine, which occurs in the form of gaseous diatomic molecules, F_2. Fluorine gas is extremely hazardous to work with, so only specially trained persons under carefully controlled conditions handle it in that form.

THE ASTROPHYSICS OF FLUORINE

The processes by which fluorine is synthesized in stars and ejected into the interstellar medium (ISM), where it is available to newly forming stars and planets, are complex and unclear. The most current understanding owes its progress to data from the NASA *Far Ultraviolet*

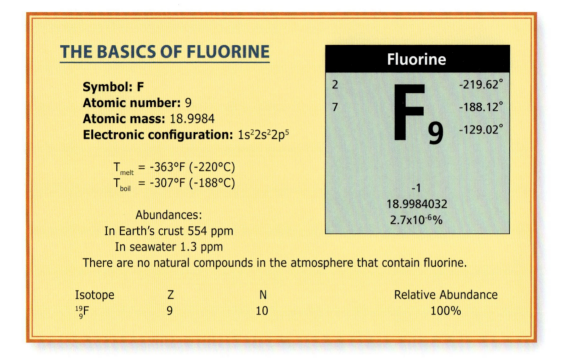

THE BASICS OF FLUORINE

Symbol: F
Atomic number: 9
Atomic mass: 18.9984
Electronic configuration: $1s^2 2s^2 2p^5$

T_{melt} = -363°F (-220°C)
T_{boil} = -307°F (-188°C)

Abundances:
In Earth's crust 554 ppm
In seawater 1.3 ppm
There are no natural compounds in the atmosphere that contain fluorine.

Isotope	Z	N	Relative Abundance
$^{19}_{9}F$	9	10	100%

Fluorine
2
7
F 9
-219.62°
-188.12°
-129.02°
-1
18.9984032
2.7×10^{-6}%

Spectroscopic Explorer (FUSE) satellite, which was commissioned in 2004 and sent data back to Earth for three years.

It has long been speculated that there are three main methods for stellar fluorine production:

1. the neutrino process in supernovae,
2. proton- and α-capture in Wolf-Rayet stars, and
3. α-capture in asymptotic giant branch (AGB) stars.

The neutrino process for production of ^{19}F (fluorine's only stable isotope) occurs when the enormous flux of neutrinos produced in a *supernova* rushes through the neon-burning shell of the exploding star. The collisions can knock a proton or neutron off ^{20}Ne. Either reaction can result in the formation of ^{19}F. Though the percentage of fluorine produced in this manner is tiny, it appears to be the main source of this element in the universe.

Smaller, but no less important, contributors to the abundance of fluorine are the Wolf-Rayet (WR) stars. These rare but massive stars, named for Charles Wolf and Georges Rayet, the Danish and American astronomers who discovered them, are unique in that they exhibit *emission line spectra.* In contrast, other stars display absorption spectra, as the surrounding gas envelope absorbs the light radiated from the inner layers of the star. It is now understood that the WR radiation comes from high-velocity gases exiting the stellar envelope prior to the supernova event that ends the life of the Wolf-Rayet star. The mass-loss mechanism is poorly understood; possibilities include thermal pulsations, radiation pressure, magnetic forces, and centrifugal forces. Whatever the forces that eject the gases, they play a crucial role in the distribution of fluorine into the ISM: any fluorine that remains in the star is lost by interactions with protons, as follows:

$$^{19}_{9}F + {}^{1}_{1}p \rightarrow {}^{4}_{2}\alpha + {}^{16}_{8}O$$

This reaction is so predominant, in fact, that fluorine *nucleosynthesis* in stars is often predicted as a function of oxygen abundance.

The third mechanism for fluorine production can occur in the asymptotic giant branch of stars, so called because of their position on the *Hertzsprung-Russell (HR) diagram*—a tool astronomers use to map the life cycles of stars. Developed independently but concurrently by the American astronomer, Henry Norris Russell (1877–1957), and Danish astronomer, Einar Hertzsprung (1873–1967), the chart plots *luminosity* as a function of temperature.

When very young stars are placed on the graph, they show up somewhere on the "main sequence" line, with the exact location depending on the star's mass. If scientists could watch one star during its entire lifetime, as it burns up its helium, becomes a *red giant,* and eventually explodes as a supernova, the star's position would move around to various points on the diagram. That, of course, is impossible to observe, as it could take hundreds of millions of years. The next best thing is to plot as many stars as possible, which give an idea of the general evolution of stars.

AGB stars map above and to the left of the red giant branch, and evolve through the red giant stage. They are intermediate mass stars $(1-7 \times M_{sun})$ whose core contains carbon and oxygen nuclei and which are characterized by thermal pulsations in the hydrogen-burning shell. These pulsations occur when a layer of helium beneath the H-shell

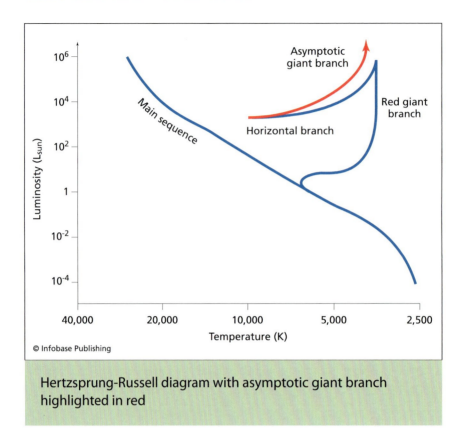

Hertzsprung-Russell diagram with asymptotic giant branch highlighted in red

begins to fuse into carbon, increasing the luminosity and engendering a high-velocity stellar wind that ejects massive amounts of stellar material, including fluorine, into space. Much controversy surrounds the question of how important the contribution of AGB stars is to fluorine abundance in the ISM, as the fraction seems to depend strongly on which particular stars are observed.

Since fluorine is so reactive, the neutral atom is difficult to detect in the ISM, but trace amounts of hydrogen fluoride have been found in interstellar clouds, though it is a difficult observation, possible only from space, where Earth's infrared frequencies can not interfere with the measurements.

THE DISCOVERY AND NAMING OF FLUORINE

The halogens are so chemically reactive that none of them is ever found in nature in pure form. As was explained previously, fluorine, chlorine, bromine, iodine, and astatine are elements that chemists classify as oxidizing agents. Oxidizing agents readily gain electrons as they react with other chemical substances. In the case of the halogens, this means that the neutral atoms become negative ions with a charge of -1 (these ions are called halide ions). In order of strength as oxidizing agents, the most reactive species is F_2, followed by Cl_2, and then Br_2, I_2, and At_2. The more reactive the pure element is, the more difficult it is to obtain the pure element from its compounds. Therefore, if the halogens were found in nature in equal quantities, we would have expected to have discovered them in the order astatine first, then iodine, bromine, and chlorine, and finally fluorine. However, they are not found in equal quantities. In fact, all of the isotopes of astatine are radioactive, with short enough half-lives that only a minute amount of astatine is found in nature, and astatine was not finally discovered until 1940.

The order of discovery of the halogens depended upon a combination of relative abundance and relative chemical reactivity. This combination resulted in chlorine being isolated first (in 1774) and fluorine being the last of the common halogens to finally be isolated (in 1886). The present story revolves around answering the question of why more than a century elapsed between the discoveries of these two elements.

The most common source of fluorine is the mineral fluorite, which has the composition CaF_2, and, as such, is 49 percent fluorine by weight. Fluorite crystals can be cubic or octahedral. The octahedral crystals are popular in rock and mineral shops, where their appeal to customers is the shape of the crystals and their colors. Some of these crystals may be quite large; they come in a variety of colors—purple, pink, green, yellow, or gray. Both the words *fluorine* and *fluorescence* were derived from this mineral. Another relatively common mineral that contains fluorine is apatite, which also contains calcium, chlorine, phosphorus, and oxygen. About 4 percent of the mineral, by weight, is fluorine and chlorine. It should be noted, too, that bone is essentially apatite, since bone has the same composition and structure that apatite does.

The fluorescence of fluorite was first noted in the 17th century. Today, scientists usually cause minerals to fluoresce by exposing them to ultraviolet (UV) light. However, ultraviolet light was not discovered until 200 years later. The discovery of the fluorescence of fluorite

Fluorine occurs naturally in the mineral fluorite (CaF$_2$). *(Bureau of Mines, Mineral Specimens)*

was made when a sample of the mineral was heated. At the same time, hydrofluoric acid (HF) was prepared from fluorite and found to etch glass, a purpose for which hydrofluoric acid is still used to this day. Of course, the composition of hydrofluoric acid remained unknown for 200 years. By the early 1800s, chemists had ruled out most of the competing candidates for the acid's composition and had concluded that the acid must contain some unknown element.

Hydrofluoric acid is poisonous, and several leading chemists suffered greatly from exposure to it. Attempts to isolate the unknown component of the acid (fluorine) proved to be even more hazardous, with several chemists dying in the attempt. (The first known fatality was Paulin Louyet, a Belgian chemist, who died in 1850 at the age of 32 while attempting to isolate fluorine gas.) Even if chemists succeeded in safely preparing a small quantity of fluorine gas, the gas would immediately react with what was around it, so obtaining pure samples remained elusive until 1886. The chemist who finally succeeded was Ferdinand-Fré-déric-Henri Moissan (1852–1907) of France.

Moissan poisoned himself several times during his first attempts to isolate fluorine gas. Finally, he dissolved potassium fluoride in liquid hydrofluoric acid, excluding water completely, and safely prepared fluorine gas. Although Moissan is best remembered for this achievement, he made numerous other contributions to science. For example, he prepared synthetic rubies. He was also the first person to prepare good-sized samples of the elements molybdenum, tantalum, niobium, chromium, manganese, vanadium, titanium, tungsten, thorium, and uranium. In addition, he prepared enough aluminum metal that his wife was one of the first women in the world to use aluminum cooking utensils. In 1906, Moissan's contributions to science were recognized when he received the Nobel Prize in chemistry. He died the following year, probably due to the cumulative effects of having worked with fluorine gas for so long.

FLUORINE GAS: A HAZARDOUS MOLECULE

With the exception of helium and neon, which do not exhibit any chemical reactivity, fluorine has the smallest atoms of any element. The small size of its atoms contributes to making fluorine the most electronegative element. In addition, the small size of fluorine atoms weakens the covalent bond in a fluorine molecule (F_2), making fluorine gas the most powerful oxidizing agent of all the elements. Because of its extreme reactivity, fluorine would never exist as F_2 in nature; fluorine is always found either bonded as the fluoride ion to other elements in salts or bonded covalently to other elements in molecular compounds. Because fluorine gas is so extremely reactive and dangerous to handle, it must be stored in special cylinders and used only by personnel trained in its use.

THE CHEMISTRY OF FLUORINE: THE MOST REACTIVE ELEMENT

Chlorine, bromine, and iodine form a great diversity of compounds because each of those elements can bond to several other atoms at a time. Fluorine is different; fluorine atoms are limited to being able to

bond to only one other atom at a time. Some of the important compounds that contain fluorine atoms are the following: hydrofluoric acid
(HF), sulfur hexafluoride (SF_6), uranium hexafluoride (UF_6), polytetrafluoroethylene (a polymer of carbon and fluorine familiar to consumers
in the form of Gore-Tex® and Teflon®), sodium and stannous fluoride
(NaF and SnF_2, respectively), and the chlorofluorocarbons (the so-
called CFCs such as CCl_2F_2.)

The properties of substances are strongly influenced by the kinds of
intermolecular forces of attraction that exist between their molecules.
Hydrogen fluoride (HF) differs significantly in its physical and chemical properties from the other *hydrogen halides* (HCl, HBr, HI). Hydrogen fluoride is a volatile liquid that boils at 20°C, which is right about
room temperature. On the other hand, the weaker forces between the
other molecules means that they vaporize more easily and have boiling
points well below room temperature. The fact that HF is a liquid rather
than a gas is attributed to *hydrogen bonding* between HF molecules that
does not occur in the other compounds. Hydrogen bonding refers to
an exceptionally strong attractive force between the hydrogen atom on
one molecule and an atom (usually nitrogen, oxygen, or fluorine) on
another molecule. Because hydrogen bonding does not occur in the
cases of chlorine, bromine, or iodine atoms, the forces of attraction
between the other hydrogen halides are weaker than the forces between
hydrogen fluoride molecules.

HCl, HBr, and HI are all strong acids, but HF is a weak acid (again,
because of the strong bond between hydrogen and fluorine). Nevertheless, HF is one of the most toxic and corrosive substances in chemistry.
It reacts chemically with glass, metals, and organic material. Hydrofluoric acid's principal use is to etch glass, but the acid must be handled
very carefully because it is easily absorbed through the skin and can
burn the skin and deeper tissues and damage bone structures. If not
treated promptly, death can result. This also means that HF cannot be
stored in glass bottles. Before the invention of plastics, HF was stored in
platinum containers. Today it is most commonly stored in polyethylene
(plastic) bottles.

Sulfur hexafluoride (SF_6) is a colorless, odorless, nontoxic gas. It is
used as a contrast agent in ultrasound imaging. In the electrical indus-

try, it serves as an insulating gas because its insulating properties are superior to air, and it can also replace the environmentally harmful *PCBs (polychlorinated biphenyls)* that the industry had used for many years. Unfortunately, SF_6 is a very potent *greenhouse gas,* and its use will probably be phased out because of increasing concerns about *global warming.* One pound of SF_6 has the warming equivalent of 11 tons of CO_2. Although the EPA currently has no mandate regarding limitations on sulfur hexafluoride emissions, some companies have taken the initiative to voluntarily reduce their SF_6 output. In February 2009, Consolidated Edison Company of New York and Arizona Public Service were commended by the EPA for their improvements in handling and maintenance, which resulted in the prevention of a quarter million pounds of SF_6 from entering the atmosphere.

Uranium hexafluoride (UF_6) was first used during the Manhattan Project (World War II, 1941–46) in the development of the atom bomb. Uranium occurs in nature as a mixture of isotopes. The most common isotope, $^{238}_{92}U$, comprises 99.3 percent of natural uranium, but it cannot be used in atom bombs or conventional nuclear power plants because it is not fissionable. The fissionable isotope—the one that can be used in power plants or weapons—is $^{235}_{92}U$, which comprises only 0.7 percent of uranium. It is not practical to try to chemically separate isotopes of the same element; on the other hand isotopes can be separated by physical processes, such as gaseous diffusion and gas centrifugation. As a sample of uranium is subjected to either process, *enrichment* takes place, meaning that the percentage of $^{235}_{92}U$ increases. When the concentration of $^{235}_{92}U$ reaches about 3–5 percent, the uranium is ready to be fashioned into fuel for a light-water nuclear reactor. Weapons-grade uranium may be as much as 85 percent $^{235}_{92}U$. UF_6 is still the form of uranium used today for isotope separation.

Teflon and Gore-Tex are common household products made from polytetrafluoroethylene, which consists of very long, linear chains of carbon atoms that have two fluorine atoms connected to each carbon. Teflon is well known for its "nonstick" properties and is often used as a nonstick coating on cookware because foods do not form any chemical bonds to it. For the same reason that Teflon does not form chemical bonds, it is also an excellent lubricant. Gore-Tex products are known for their water-resistant qualities. Outdoor clothing and footwear are often lined with Gore-Tex to protect the wearer from wet conditions.

Fluorine commonly occurs in compounds as the fluoride ion (F^-) in salts like sodium fluoride (NaF) and stannous fluoride (SnF_2). Both of these compounds may be added to toothpaste to fight cavities. Although fluoride itself is effective in the fight against tooth decay, several precautions must be taken. Too much fluoride will result in mottling of the teeth. Some people are sensitive to table salt (sodium chloride, NaCl) in their diet, and sodium fluoride could cause the same problems. Stannous fluoride may cause yellowing of the teeth. Because of these potential problems, most major toothpaste companies offer consumers choices of toothpaste with or without fluoride.

Chlorofluorocarbons were originally developed as refrigerants and sold under the commercial name Freon®. CFCs were used for decades in refrigerators and automobile air-conditioning units. However, in the 1980s, the chlorine in CFCs was identified as a substance harmful to the layer of ozone in the stratosphere that protects life on Earth from the Sun's harmful ultraviolet rays. Since then, CFCs have been gradually replaced with HFCs, hydrofluorocarbons that do not contain chlorine.

FLUORIDE IN DRINKING WATER: THE DEBATE

Today, more than half of U.S. water supplies is supplemented with fluoride at a level between 0.7 and 1.2 ppm. This practice originated when fluorine, quite by accident, was discovered to promote stronger teeth.

In the early part of the last century, a dentist named Frederick McKay investigated possible causes of tooth discoloration (now called "dental fluorosis") in a certain population of Colorado children. He also noticed that those children had fewer cavities than average. The reason for both phenomena turned out to be a higher than average concentration of fluoride ions in the drinking water—a result of leaching of the mineral cryolite into the local reservoir. Dr. McKay's research eventually came to the attention of many scientists and local water authorities, who began to add fluoride compounds to drinking water. Toothpaste companies decided a good market move would be to increase fluoride availability. Most brands of toothpaste now include fluoride in unspecified amounts.

Many brands of toothpaste contain fluoride. *(Tobi Zausner)*

New studies, however, show that the advantages of fluoride enhancement may not outweigh the risks. Fluorine tends to accumulate in bones, joints in particular, which has contributed to the threefold increase in hip fractures over the last century. Fluoride compounds have also tested positive as carcinogens, and there may be reproductive risks. Ethical issues are also coming to the forefront: Is it right to mandate medication in any form?

TECHNOLOGY AND CURRENT USES

Because of the extreme hazard of handling fluorine gas, relatively little use is made of it. In the form of a fluoride ion and in compounds, however, it has several important uses.

In the interests of dental health, fluoridation of municipal drinking water supplies has been practiced throughout the United States for several decades, though the practice has recently come under much criticism. Likewise, fluoride—as either NaF or SnF_2—is often added to toothpaste to fight tooth decay. Fluoride compounds have also tested positive as carcinogens, and they may cause reproductive risks.

The compound hydrofluoric acid is used to etch glass, and fluorine-18 is used in positron emission topography (PET)—a nuclear medicine imaging technique. In addition, fluorine compounds have been used as refrigerants in air conditioners, refrigerators, and freezers. These compounds, however, are in the process of being phased out, according to Montreal Protocol restrictions that are intended to protect the ozone layer in Earth's stratosphere. (See chapter 3.)

Several synthetic materials are made of fluorinated hydrocarbons. Teflon and Gore-Tex are common household products made from polytetrafluoroethylene, which consists of very long linear chains of carbon atoms that have two fluorine atoms connected to each carbon. Gore-Tex products are known for their water-resistant qualities. Outdoor clothing and footwear are often lined with Gore-Tex to protect the wearer from wet conditions. Teflon is well known for its nonstick properties and is often used as a nonstick coating on cookware because foods do not form any chemical bonds to it. Another Teflon component, perfluorooctanoic acid, has recently come under scrutiny as a possible carcinogen, and in 2006 the EPA recommended a 95 percent phaseout of its use by 2010.

Uranium hexafluoride (UF_6) was first used as part of the Manhattan Project (1942–46) in the development of the atom bomb during World War II. UF_6 is still the form of uranium used today for isotope separation.

2

Chlorine: From Table Salt to Safe Swimming

Chlorine is element number 17. While chlorine is abundant in seawater and in Earth's crust, it is only the 20th most abundant element. Chlorine is most commonly found in nature in the form of the chloride ion (Cl^-) in salts. So-called table salt is sodium chloride (NaCl), which has been used since prehistoric times to flavor food. In fact, sodium chloride is such an important *electrolyte* for all living organisms that animals are known to frequent outcroppings of sodium chloride called salt licks to satisfy their cravings for salt.

In its pure form, chlorine is an odorous, pale green gas in the form of gaseous diatomic molecules with the formula Cl_2. Because chlorine gas and other forms of chlorine are such powerful oxidizing agents, chlorine is used in many countries around the world to disinfect municipal drinking water supplies and to treat the water in swimming pools. The

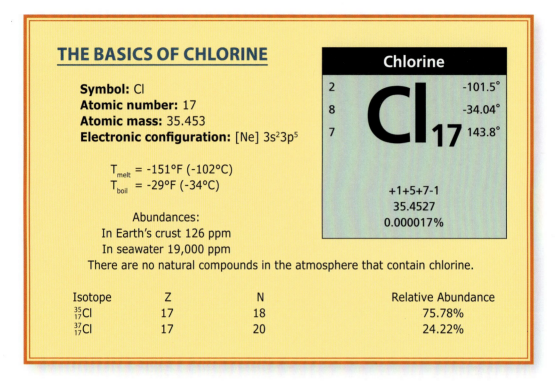

THE BASICS OF CHLORINE

Symbol: Cl
Atomic number: 17
Atomic mass: 35.453
Electronic configuration: [Ne] $3s^2 3p^5$

T_{melt} = -151°F (-102°C)
T_{boil} = -29°F (-34°C)

Abundances:
In Earth's crust 126 ppm
In seawater 19,000 ppm
There are no natural compounds in the atmosphere that contain chlorine.

Isotope	Z	N	Relative Abundance
$^{35}_{17}$Cl	17	18	75.78%
$^{37}_{17}$Cl	17	20	24.22%

Chlorine

2
8
7

Cl 17

-101.5°
-34.04°
143.8°

+1+5+7-1
35.4527
0.000017%

introduction of chlorine in the 19th century to disinfect water has been extremely beneficial to humankind by killing harmful bacteria like *E. coli* and in the prevention of waterborne infectious diseases like cholera and giardia. In addition, chlorine is found in numerous household products in the form of disinfectants and bleach.

THE ASTROPHYSICS OF CHLORINE

Chlorine atoms, particularly the isotopes ^{35}Cl and ^{37}Cl, are synthesized during explosive oxygen burning within supernovae, which are common, but rarely observed. More common are nova explosions, which occur in two-star or "binary" systems. It is worth noting that nearly 80 percent of stars in the universe belong to such systems in which one star orbits another. In some, a massive yet compact white dwarf star attracts hydrogen gas from the outer envelope of its companion star, which could be an ordinary star like the Sun. Because of the white dwarf's extreme mass, the accretion of material is so gravitationally energetic that temperatures quickly reach fusion capability, which triggers an explosive event that is detected by astronomers as a sudden brightening of the normally dim dwarf star. Unlike the case with supernovae, the explosion does not destroy the star, but merely blows the fused gases into space. In the chaos of the explosion, many elements, including chlorine, are synthesized and ejected into the interstellar medium (ISM).

Astronomers look to spectra from novae, but also other sources for information on chlorine in the universe. The abundance of chlorine produced in the universe can be studied by examining the spectra of the unfortunately named "planetary nebulae," which have nothing to do with planets, but were once thought to have some such association. These so-called planetary nebulae come about when some asymptotic giant branch (AGB) stars eject part of their atmospheres into the ISM. Because chlorine is probably not produced in this type of star, levels of the gas observed in these stellar ejections can inform scientists about chlorine concentrations that existed before the star formed. Chlorine

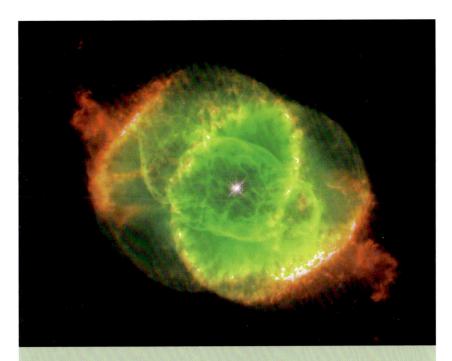

Planetary nebulae, like the beautiful example "NGC 6543" shown here, can help astronomers understand chlorine concentrations in interstellar space. (*Hubble Space Telescope Center*)

spectra are especially useful as observable markers for the abundance of cold hydrogen gas in the ISM because of the positive chlorine ion's ease of reaction with H_2.

Chlorine ions have also been detected in the atmosphere of Jupiter's moon, Io. This small moon is not massive enough for gravity to hold on to its entire atmosphere, so much of it escapes into space. It cannot go far, however. Because of collisions among molecules and interactions with sunlight, electrons become detached, resulting in ions. These are easily trapped by Jupiter's strong magnetic field, which cannot affect neutral species, but can hang on to charged atoms and molecules. The ions are distributed all along Io's orbital path around the mother planet, forming a torus or doughnut of ions, described as a plasma. Io's plasma torus displays an unexpectedly high chlorine abundance—2 percent of all ions in the ring, a concentration higher than observed in any other planet's atmosphere. Sources may include HCl, SCl_2, and Cl_2 gases, which are known constituents of volcanic eruptions on Earth. Some scientists, however, argue that NaCl (common table salt) is a good candidate. NaCl could exist as solid deposits on Io's surface or in gaseous form in the atmosphere. It is an exciting area of current research and discussion among planet scientists.

THE NOVELTY OF THE NEGATIVE ION

The negative chlorine ion can be stimulated to a state where two electrons simultaneously jump to a higher quantum state. The electrons are correlated in that they do not move independently, but act as an interlocked pair during the excitation. As with the negative hydrogen ion, when the ion-relaxation process occurs, one electron loses energy and jumps back to a lower state while the other absorbs that energy— enough to allow it to escape from the attraction of the positively charged nucleus. The same behavior has been observed in many atoms and ions that have only two electrons in their outer shells.

Negative chlorine is unique in that it is the only ion (or atom) for which a six-electron correlated state has been induced. To create this state, scientists used *synchrotron radiation* produced by the Advanced Light Source at Lawrence-Berkeley National Laboratory. Photons with a range of energies were aimed at the negative chlorine ion target. At photon energies above 40 electron volts (the energy that would simply detach two electrons, leaving a residual Cl^+ ion), they detected Cl^{2+} ions, which meant that three electrons must have been detached. The process by which this happens is complex and theoretically difficult, but it is probable that it involves a highly correlated motion of all six electrons in the outer (3p) subshell. Most likely, only one electron absorbs the photon and must somehow release that energy. One might think it would jump to a higher state and then decay, but it does not. Instead, it seems to distribute the energy to the other electrons in the closed subshell. They share the commotion, and three of them leave at once. It is an interesting process, but one that may never be understood, as the calculations are so complex as to be perhaps unattainable.

THE DISCOVERY AND NAMING OF CHLORINE

Rock salt (NaCl) has been used to flavor and preserve food since prehistoric times. Salt was a valuable commodity in ancient times and—before the invention of coinage—was used as a medium of exchange in bartering. The first reference to salt in the Bible (and also the first reference to sulfur, or "brimstone") was recorded in the book of Genesis, chapter 19, when "Lot's wife behind him looked back, and she became a pillar of salt." As an example of the high value attributed to salt, in the teachings referred to as the Beatitudes (found in the Gospel of Matthew, chapter 5), Jesus is reported to have used the following analogy when he said to his followers, "You are the salt of the earth; but if salt has lost its taste, how shall its saltiness be restored? It is no longer good for anything except to be thrown out and trodden under foot by men" (New Revised Standard Version).

It was not until 1774, however, that German-Swedish chemist Carl Wilhelm Scheele (1742–86) isolated pure chlorine as an element in the form of chlorine gas (Cl_2). This discovery was actually incidental to Scheele's main purpose. Several chemists had attempted unsuccessfully to isolate metallic manganese as the pure element. Scheele attacked the problem by dissolving the mineral pyrolusite (MnO_2) in hydrochloric acid (HCl), which at that time was called marine acid (and today is also called muriatic acid). Scheele failed to isolate manganese. However, his experiments paved the way for Swedish chemist and metallurgist Johan Gottlieb Gahn (1745–1818) to succeed in doing so later that same year.

Although Scheele failed to isolate manganese, he did make note of the choking green gas that was produced by the reaction between MnO_2 and HCl. Scheele called this gas "dephlogisticated marine gas," since he thought it resulted from the removal of a hypothetical substance called phlogiston from marine gas. Scheele observed that chlorine acts as a bleach and that it attacks all metals.

Later, in 1785, French chemist Claude-Louis Berthollet (1748–1822) noticed that aqueous solutions of chlorine had the same bleaching effect as chlorine gas itself. Berthollet's report generated enough interest among persons in the textile industry that the Scottish engineer James Watt (1736–1819)—best known as the inventor of the steam engine—bleached 1,500 yards (1,371.6 m) of linen. Watt's success prompted English doctor, Thomas Henry (1734–1816), to establish commercial bleaching operations in the city of Manchester. Unfortunately, working directly with either chlorine gas or chlorine water produces harmful effects on the respiratory system. Therefore, people working in this field sought other substances that contain chlorine, but which would be safer to handle. Bleaching powders were developed, including substances like calcium hypochlorite [$Ca(OCl)_2$] and sodium hypochlorite (NaOCl, the ingredient today in common household bleach). With the ability to bleach cotton, the use of cotton in clothing began to grow.

By the early 1800s, chlorine was also recognized as a disinfecting agent. Most public water supplies today are disinfected with chlorine agents.

Scheele is credited with the discovery of chlorine, although he believed it to be a compound, not a pure element. Today, the reaction between MnO_2 and HCl is still used in laboratories to prepare chlorine gas. Chlorine—and sodium metal—also are prepared today commercially by the electrolysis of molten sodium chlorine obtained either in the form of rock salt or from the evaporation of seawater. The chemical equation for the reaction is the following:

$$2 \text{ NaCl (aq)} \rightarrow 2 \text{ Na (s)} + Cl_2 \text{ (g)}.$$

In 1810, English chemist Sir Humphrey Davy (1778–1829) concluded that chlorine was indeed an element and not a compound. It was Davy who suggested the name "chlorine" from the Greek word for "green."

THE CHEMISTRY OF CHLORINE

With the exception of some of the noble gases, probably every element in the periodic table forms compounds with chlorine. Metal ions with +1 or +2 charges, such as Na^+, Ag^+, Ca^{2+}, and Fe^{2+}, form ionic compounds with the chloride ion. Metal atoms in higher oxidation states form molecular compounds with chlorine atoms; examples include $FeCl_3$, $TiCl_4$, and $AlCl_3$. Metalloids in positive oxidation states also readily form molecular compounds with chlorine; examples include BCl_3 (also written as B_2Cl_6), $SiCl_4$, $AsCl_3$, and $SbCl_3$. Nonmetals are capable of forming molecular compounds with chlorine in which the nonmetal could be in either a positive or a negative oxidation state. Examples in which the nonmetal is in a positive oxidation state (and chlorine is in the -1 state) include CCl_4, NCl_3, SCl_2, and $BrCl$. Examples in which the nonmetal is in a negative oxidation state (and chlorine is in a positive state) mostly involve compounds with oxygen, and include Cl_2O, ClO_2, $HClO_3$, and $HOCl$.

Chlorine has a more varied chemistry than fluorine does. Chlorine's chemistry is more diverse largely because chlorine atoms are larger than fluorine atoms; chlorine atoms have almost twice as many electrons; and the electrons in chlorine atoms can be promoted to d orbitals that permit more bonding positions. Hence, a chlorine atom can bond to multiple other atoms, whereas a fluorine atom can bond to only one other atom. As stated previously, chlorine is best known for its power as an oxidizing agent. However, unlike fluorine, which is an oxidizing agent mainly as elemental F_2, chlorine has a number of common forms in which it is an oxidizing agent—chlorine gas (Cl_2); the salts $NaClO$, $NaClO_2$, $NaClO_3$, $NaClO_4$, and $Ca(ClO)_2$; the acids $HClO$, $HClO_2$, $HClO_3$, and $HClO_4$; and the compound ClO_2.

Chlorine can occur in a large number of oxidation states: -1, 0, +1, +3, +5, and +7. Therefore, chlorine is an example of an element that can undergo *disproportionation*, or autooxidadtion, reactions. In disproportionation, one substance serves as both the oxidizing agent and as the reducing agent. With chlorine, an example of a disproportionation reaction is the following:

$$3 \ Cl_2 + 6 \ KOH \rightarrow 5 \ KCl + KClO_3 + 3 \ H_2O,$$

in which chlorine begins in the 0, or neutral, oxidation state. Some chlorine atoms are oxidized to the +5 state (in $KClO_3$), while the remaining atoms are reduced to the -1 state (in KCl).

Reactions can also occur between compounds in which chlorine begins in different oxidation states, and those states change. For example, in the reaction

$$2 \ HCl + NaClO \rightarrow Cl_2 + NaCl + H_2O,$$

chlorine atoms begin in the -1 and +1 states and are changed to the 0 state in Cl_2. This reaction demonstrates the hazards of mixing common household products without knowing what the consequences might be. Vinegar contains acetic acid, which is not as strong as hydrochloric acid, but which is an acid nevertheless. Bleach contains NaClO. Therefore, mixing the two together at home could produce noxious chlorine gas. (An even more common mistake, and potentially more deadly, is made by mixing together bleach and household ammonia. One should never use the two cleaners together; they will react to produce a very toxic gas that can be deadly in an enclosed environment such as a bathroom. Several deaths have resulted from persons making this mistake.)

Since chlorine gas is poisonous and therefore dangerous to transport or store on-site, chlorine in compound form is more likely to be used in bleach and to purify or disinfect water supplies. Common household bleach, such as Clorox®, is a solution of 5 percent sodium hypochlorite (NaClO). The introduction of chlorine to disinfect municipal water supplies has been one of history's great public-health success stories. Chlorination has substantially reduced the rates of typhoid, cholera, hepatitis, and giardiasis. The most common compounds used are sodium hypochlorite, calcium hypochlorite [$Ca(OCl)_2$], and chlorine dioxide (ClO_2). Other disinfectants such as ozone (O_3) may be used, but a disadvantage of ozone is that there is no residual effect. In other words, even though most of the pathogens are destroyed upon contact with ozone, no residual ozone remains, so that a small number of pathogens survive, and their numbers can rebound. With chlorine, however, residual chlorine remains in the water, inhibiting the rebound of pathogens.

Naturally occurring inorganic compounds that contain chlorine do so mostly in the form of chloride ions (Cl^-). One of the most common chemical compounds is table salt—sodium chloride, or $NaCl$—obtained either by evaporation of seawater or by mining the mineral halite (also known as rock salt). With few exceptions ($AgCl$, $PbCl_2$, and Hg_2Cl_2), all compounds that contain chloride ions tend to be very soluble in water. Consequently, there tend to be very few terrestrial sources of chloride compounds other than rock salt. In addition, there tends to be very little chemistry of the chloride ion itself. Even in hydrochloric acid (HCl), which is an industrially important reagent that contains the chloride ion, the chemistry is mostly of the hydrogen ion (H^+), not the chloride ion.

Rock salt crystals grow in many shapes and shades. *(Craig Barhorst/Shutterstock)*

Most people know of chlorine because of its infamous nature. Chlorinated hydrocarbons were first developed in the first half of the 20th century for uses such as solvents, dry-cleaning agents, insecticides, plastics, refrigerants, and electrical insulators. In addition, a large fraction of the chlorinated hydrocarbons found in nature is produced by metabolic processes in plants and by natural events such as volcanic eruptions. Unfortunately, most of these substances are extremely persistent, meaning that they are not easily broken down in the environment, but remain in the environment for several years and often accumulate in the food chain.

Many of these compounds are carcinogenic, harmful in other ways to human beings or damaging to wildlife. An especially notorious example of a pesticide is dichlorodiphenyltrichloroethane (DDT), which was introduced because mass application is easy and—in low doses at least—is relatively nontoxic to humans. These properties make DDT a cost-effective weapon for combating insect-borne diseases such as malaria. Unfortunately, DDT does not readily break down in the environment into harmless compounds, but instead accumulates in the tissues of organisms. As DDT works its way up the food chain, its concentration in animal tissues increases to the point where it disrupts reproduction in vertebrate species. (DDT's harmful effects were especially well documented in Rachel Carson's 1962 environmental classic, *Silent Spring*.)

Several common plastics are manufactured using chlorine. The most important example is polyvinyl chloride (PVC). PVC has many uses, including applications in plumbing, vinyl siding, magnetic stripe cards, upholstery, and flooring.

Both the qualitative detection of chloride ion and the quantitative analysis of chloride in a solution are relatively simple. The chloride salts of most metals are soluble in water, AgCl being a notable exception. Silver chloride is a white solid that is easily soluble in dilute solutions of aqueous ammonia, as shown in the following reaction:

$$\text{AgCl (s)} + 2\,\text{NH}_3\,\text{(aq)} \rightarrow \text{Ag(NH}_3)_2^+\,\text{(aq)} + \text{Cl}^-.$$

Silver also precipitates with bromide and iodide ions. However, silver bromide dissolves only in concentrated aqueous ammonia, and silver iodide does not dissolve at all. A chloride solution can be titrated with silver nitrate using potassium chromate (K_2CrO_4) as an end-point indicator. Reddish brown silver chromate (Ag_2CrO_4) is less soluble than silver chloride, so silver chloride precipitates first. When all of the chloride has precipitated, silver chromate then precipitates, as shown in the following reaction:

$$2 AgNO_3 (aq) + K_2CrO_4 (aq) \rightarrow Ag_2CrO_4 (s) + 2 KNO_3 (aq).$$

As Ag_2CrO_4 precipitates, the mixture turns from white to reddish brown. The amount of $AgNO_3$ that was added is used to calculate the concentration of chloride ion in the original solution.

FIREWORKS AND FLAME RETARDANTS

Fireworks require several key ingredients; for example, they must contain chemicals that can react together to produce high temperatures, chemicals that can emit light, and chemicals that can give the light various colors. Chlorine-containing compounds help satisfy two of these requirements.

Chlorates (compounds containing the ClO_3^- ion) and perchlorates (compounds containing the ClO_4^- ion) are powerful oxidizing agents. When these compounds react with powerful reducing agents such as sulfur or charcoal, not only can high temperatures be reached, but the reactions can be explosive in nature.

Chlorine-containing compounds also contribute to giving fireworks their spectacular colors. *Luminescent* substances produce light by transitions of electrons between energy levels. Examples of chlorine-containing compounds include barium chloride, which gives a green color, and copper chloride, which gives a blue color. Other colors such as reds and oranges may be produced by various metals in the form of chloride salts.

Compounds that are high in chlorine or bromine are flame resistant. Therefore, flame retardants are incorporated into clothing, especially into clothing designed for children and clothing worn by firefighters. Some camping materials, such as tenting fabric, also contain flame retardants.

(continues on page 34)

CHLORINE AS A WEAPON OF WAR

Chlorine's greatest claim to infamy was probably its use in chemical warfare in the trenches of World War I. Chlorine gas irritates mucous membranes in eyes, nose, and throat. It has been described as a blistering agent. Upon entering the lungs, it can ravage lung tissue by reacting with water vapor to form hydrochloric acid, noted as follows:

$$2Cl_2 + 2\,H_2O \rightarrow 4\,HCl + O_2.$$

There is also the possibility of asphyxiation. Death is likely when the concentration in air approaches 1,000 ppm.

The French were the first to realize the viability of chlorine gas as a weapon. Easily acquired as a by-product of dye produc-

Chlorine gas was used as a chemical weapon in World War I. *(Library of Congress)*

tion, it could be stored in cylinders and released when a wind likely to carry it across enemy lines presented itself. Naturally, that turned out to be an important limitation. In addition, the gas was clearly visible—a greenish cloud drifting toward the enemy, who could take evasive action. Those actions included running through (not away from) the cloud, thereby minimizing the time spent in it, and placing a wet cloth over the face, which neutralized the effect—perhaps catalyzing the above HCl reaction in the cloth rather than in the lungs.

Despite its limitations, Cl_2 was an inexpensive weapon that gained favor with the German military machine, resulting in large-scale assaults against Britain, which retaliated reluctantly but with vigor. British gas attacks, in fact, grew to be more numerous than Germany's during the last two years of the war. This might seem surprising, since soldiers could fairly easily avoid or minimize their exposure once they perceived the threat. The reason for its continued use was its effect as a powerful psychological tool. Fear was invoked in those who had experienced or heard how painful the effects of the pure gas and its compounds (especially the notorious mustard gas) could be. In the end, however, the gas was responsible for only a minuscule fraction of deaths in the war.

After the war, the *Geneva Protocol* ruled out the use of gas in warfare as "justly condemned by the general opinion of the civilised world." Unfortunately, there are still parties of insurgents in war-torn areas like Iraq who find chlorine gas to be a convenient and effective tool of war. These groups are differently motivated than the British and German military. If the purpose is to kill, injure, or frighten as many people as possible, chlorine gas, which is easily attainable, serves them well. Vehicles that are filled with chlorine gas cylinders and then detonated can spread panic and disorder and do not rely on wind direction. Just such a plan carried out in March 2007 killed two and sickened as many as 350 Iraqi citizens.

(continued from page 31)

TECHNOLOGY AND CURRENT USES

Chlorination is the most common method employed by municipalities in the United States to disinfect potable water. The use of chlorine to disinfect municipal water supplies has been one of history's great public health success stories. Chlorination has substantially reduced the rates of typhoid, cholera, hepatitis, and giardiasis. The most common compounds used as disinfectants are sodium hypochlorite, calcium hypochlorite, and chlorine dioxide.

Chlorine is also an important element in a large number of useful organic compounds. For example, chlorine is added to hydrocarbon chains in the manufacture of plastics and fabrics, the most important example being polyvinyl chloride (PVC). PVC has many uses, including applications in plumbing, vinyl siding, magnetic stripe cards, upholstery, and flooring. Many medicines and other pharmaceuticals are manufactured in the form of chloride compounds. Chlorinated hydrocarbons represent a large class of insecticides. Chlorine's bleaching action is also well known. Common household bleach is a solution of five percent sodium hypochlorite.

In other uses, chlorinated, *nonpolar solvents* include carbon tetrachloride and chloroform. Chlorine-containing substances are responsible for some of the vivid colors in fireworks: Luminescent substances produce light by transitions of electrons between energy levels. Examples of such chlorine-containing compounds include barium chloride, which gives a green color, and copper chloride, which gives a blue color. Other colors, such as reds and oranges, may be produced by various metals in the form of chloride salts.

Compounds that are high in chlorine or bromine are flame-resistant. Therefore, flame retardants are incorporated into clothing, especially into clothing designed for children and clothing worn by firefighters. Some camping materials, such as tenting fabric, also contain flame retardants.

3

Bromine: Unusual at Room Temperature

Bromine, element 35, is the 47th most abundant element in Earth's crust. Like chlorine, bromine is more abundant in seawater than in terrestrial rocks and minerals. However, unlike fluorine and chlorine, which in the form of pure elements are gases under normal atmospheric conditions, bromine is a liquid. In fact, bromine is one of only two elements in the periodic table that is a liquid at atmospheric pressure and room temperature, the other example being mercury. Liquid bromine is an odorous, volatile, deep red, viscous, toxic, and corrosive liquid that is very hazardous to handle. In contact with the skin, bromine causes instant injury. It is especially important to avoid inhaling the fumes. In fact, bromine's fumes tend to be more dangerous than chlorine's fumes because bromine is denser and can settle more deeply into the lungs.

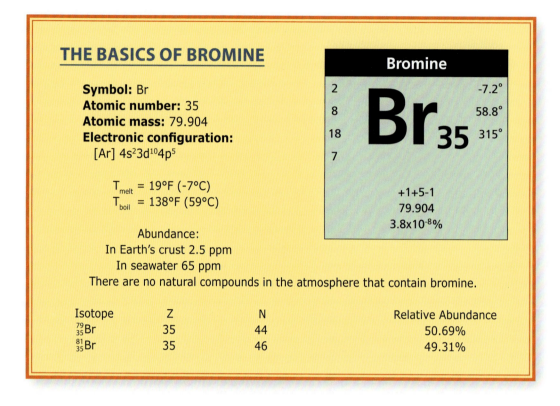

THE BASICS OF BROMINE

Symbol: Br
Atomic number: 35
Atomic mass: 79.904
Electronic configuration:
  [Ar] $4s^23d^{10}4p^5$

  T_{melt} = 19°F (-7°C)
  T_{boil} = 138°F (59°C)

Abundance:
In Earth's crust 2.5 ppm
In seawater 65 ppm
There are no natural compounds in the atmosphere that contain bromine.

Bromine

2
8
18
7

Br 35

-7.2°
58.8°
315°

+1+5-1
79.904
3.8x10^{-8}%

Isotope	Z	N	Relative Abundance
$^{79}_{35}$Br	35	44	50.69%
$^{81}_{35}$Br	35	46	49.31%

Bromine's chemistry is very similar to chlorine's chemistry. Bromine is an oxidizing agent, although not as powerful an oxidizing agent as chlorine or fluorine. In addition, bromine forms numerous chemical compounds that are analogs of the corresponding chlorine compounds.

THE ASTROPHYSICS OF BROMINE

Bromine is a curiosity in astronomy because it is so rarely observed in stellar atmospheres. When it is observed, its existence is difficult to explain. Neutral bromine (Br I) as well as the positive ion (Br$^+$ or Br II) have all been detected in a few "chemically peculiar" stars, so named because of their chemical peculiarity. These stars show anomalies in the expected abundance of various elements (usually as compared to solar abundance). They have been categorized according to the uncommon nature of their spectra. Some have extra helium or a strange mercury-manganese combination. Bromine is more often found in so-called hot peculiar stars that have high temperatures and unusual surface compositions, and appear on the *main sequence* of the Hertzsprung-Russell

(HR) diagram. Bromine is also observed in the spectra of particular white dwarf stars—those with helium-rich atmospheres.

The observations have been difficult for astrophysicists to explain because it was previously believed that any element heavier than iron must have been synthesized in *supernova* events that blow the gases into space rather than retaining them in the stellar atmosphere.

Details on bromine nucleosynthesis in these stars are, therefore, tentative at best. Commonalities do not exist in the data. This means that observations are believable on an individual basis. No particular class of star exhibits bromine in its atmospheric spectra.

Possible avenues of nucleosynthesis include convective heat waves (or thermal pulsations) that allow the mixing of elements needed to guarantee creation of heavier elements. Stellar winds, radiative effects, turbulence, convection, and accretion from stellar companions may also play a part. Unfortunately, there are insufficient observations to clarify the causes. Fortunately for future astrophysicists, there is still a mystery to be explored.

THE DISCOVERY AND NAMING OF BROMINE

Sea salt is a mixture of soluble ionic compounds. The positive ions are mostly sodium and potassium, and the negative ions are mostly chloride and iodide. However, there is also some bromide ion, and it was from sea salt that the element bromine was first isolated.

Bromine was isolated independently by two chemists—Antoine-Jerome Balard (1802–76) in France and Carl Lowig (1803–90) in Germany. Lowig had actually prepared bromine first, but Balard published his findings first, so Balard is usually credited as the discoverer.

Balard was a professor of chemistry at the Sorbonne and at the College de France, which are both located in Paris. Balard was interested mostly in pursuing work of economic importance. It was the year 1826. Iodine had been discovered in 1811, and already had been prescribed in 1820 as a treatment for *goiter,* an enlargement of the thyroid gland. Balard enjoyed being near the ocean, and was investigating ways of extracting iodine from sea water (for example, with ether) when he noticed that, in addition to iodine, he was obtaining a liquid that was yellow in solution. Upon distilling the liquid in order to purify it, Balard

Bromine's fumes can be more dangerous than chlorine's. *(Charles D. Winters/Photo Researchers, Inc.)*

obtained a smelly dark-red liquid that he wanted to call "muride," but which other chemists convinced him to call "bromine," from the Greek word meaning "stink."

In the meantime, Lowig had already prepared bromine but was still studying it when Balard's discovery was announced. In fact, several chemists read Balard's report and realized that they, too, had prepared bromine, but they had not recognized it as a new element (they thought it was a form of chlorine). It was to Balard's credit that bromine was in fact recognized as a new element that shared properties similar to the properties of chlorine and iodine.

THE CHEMISTRY OF BROMINE

Chlorine and iodine are named for their colors. In its pure elemental form, bromine is a dark-red, volatile liquid with a harsh stench that gives bromine its name. Pure bromine is very hazardous to handle; it burns the skin and its red, heavy vapors, if breathed, are toxic.

In the United States, most bromine is obtained by its recovery from seawater or from underground salt mines left behind by prehistoric oceans. There is approximately 1 pound (0.454 kg) of bromine in every 1,800 gallons (6,813.7 liters) of seawater. The major current use of bromine is in the synthesis of organic compounds.

The chemistry of bromine is similar to the chemistry of chlorine, especially because both elements can exist in the same large variety of oxidation states (-1, 0, +1, +3, +5, and +7). In contrast to chlorine, however, in which compounds in all of these oxidation states are common, the most common bromine compounds occur only in the -1 and +5 states. Bromine compounds in the other oxidation states have been made, but they tend not to have any practical uses.

Both inorganic and organic bromide compounds are used in industry as *brominating agents*. In other words, they are used to add bromine atoms to other compounds. For example, among inorganic substances, there are the compounds phosphorus tribromide (PBr_3), bromine chloride (BrCl), aluminum bromide ($AlBr_3$), ammonium bromide (NH_4Br), and thionyl bromide ($SOBr_2$). Organic syntheses usually proceed by a series of stepwise reactions in which intermediate substances are made in one step and then converted into something else in the next step. Adding bromine is common because bromine is both easily added and easily removed. Chemists often add bromine atoms to carbon chains containing double or triple bonds to make intermediates that can then be converted to the desired end products. An example is shown in the following equation:

$$CH_2 = CH_2 \ (aq) + Br_2 \ (aq) \rightarrow CH_2Br—CH_2Br \ (aq).$$

A common class of brominating compounds consists of the alkyl bromides, where an alkyl group is a hydrocarbon fragment such as $CH_3\cdot$, $CH_3CH_2\cdot$, or $CH_3CH_2CH_2\cdot$ (the dot [·] is the symbol for an unpaired electron; the dot is common for fragments of molecules). For example, if the alkyl group is a methyl group ($CH_3\cdot$), the brominating agent would be methyl bromide (CH_3Br).

The acid of bromide salts is hydrobromic acid (HBr), which is a strong acid. However, its use is much less common than that of the strong acid, hydrochloric acid (HCl). Although many chlorine-

Reaction between chlorine and bromide ions *(Charles D. Winters/ Photo Researchers, Inc.)*

containing compounds have bromine analogs, chlorine is less expensive to produce and—despite chlorine's own hazards—is safer to handle than bromine is. Therefore, there tends to be very little demand for the bromine-containing compounds.

The qualitative analysis of bromide ions in solution relies on the fact that silver bromide is intermediate in solubility between silver chloride and silver iodide. Silver ions precipitate with a number of negative ions in addition to the halides, ions that include PO_4^{3-}, CO_3^{2-}, and S^{2-}. Acidifying an unknown solution first with nitric acid (HNO_3), however, removes the other ions, leaving the halides. As described in the section on "The Chemistry of Chlorine" earlier in this chapter, silver chloride (AgCl) dissolves readily in dilute aqueous ammonia, whereas silver bromide (AgBr) dissolves only in concentrated aqueous ammonia (and silver iodide [AgI] does not dissolve at all in ammonia). Therefore, a definitive test for the presence of bromide ion in a solution is to acidify the solution with nitric acid and then add silver nitrate. If a precipitate forms, it is most likely to be one of the halides. If the precipitate is insoluble in dilute ammonia, but soluble in concentrated ammonia, then the presence of bromide ions in the solution is confirmed.

These reactions are shown in the following equations:

$$AgNO_3 \text{ (aq)} + KBr \text{ (aq)} \rightarrow AgBr \text{ (s)} + KNO_3 \text{ (aq)};$$

$$AgBr \text{ (s)} + 2\,NH_3 \text{ (aq)} \rightarrow Ag(NH_3)_2^+ \text{ (aq)} + Br^- \text{(aq)}.$$

On the other hand, if the solid is yellow in color and does not dissolve in concentrated ammonia, it is most likely AgI, and it may be concluded that the original unknown solution contained iodide ions.

A RADIOACTIVE NATURE

Bromine has a significant number of radioactive isotopes as well as a couple of *isomeric states* that make it a continuing source of interest in the nuclear physics research community. Most of bromine's radioactive isotopes decay by emitting high-energy photons (gamma rays), electrons (beta particles), or neutrons. Some, however, emit positrons. A positron is the *antiparticle* to the electron. As such, it has the same properties as the electron—such as mass, radius, and spin—but opposite charge. A positron colliding with an electron meets a strange fate. Both particles disappear or "annihilate," and their equivalent energy is shared by two photons that are emitted in opposite directions from the collision vector.

Positrons are useful in medicine because electron current can measure the brain's activity. A positron emitter, such as ^{75}Br, injected into a person's bloodstream, will eventually reach all areas of the body, including the brain. The brain does not normally produce high-energy photons. If they are detected, it means positrons met with electrons in a high-activity area. This is the basis of the PET-scan. PET stands for positron emission *tomography*. A tomogram thus produced gives a two-dimensional picture of a very particular region or slice of the brain. A PET-scan can look at a multitude of neighboring slices to find possible indications of edema, Parkinson's disease, or lung cancer. In a 2001 *Scientific American* article, Dr. Mony J. de Leon, of the New York University School of Medicine, comments that the use of PET-scans may even help "find a way to delay the onset of Alzheimer's or prevent it altogether."

Bromine also displays interesting metastable isomeric states. These are excited states of a nucleus different from, but similar to, excited states of an atom. An atom is in an excited state if an electron absorbs energy and jumps to a higher orbit. A nucleus is in an excited state if a nucleon absorbs energy and changes its spin. In either case, the energy to force the change is delivered by collisions with photons or massive particles like protons, neutrons, deuterons, or alpha particles.

Bromine's isomers are somewhat mysterious. Bromine 80 can maintain its state for relatively long (4.5-hour) or short (18-minute) periods. Only the latter involves a gamma ray decay product. Bromine 83, on the other hand, has a half-life of about 2.4 hours, but decays via two successive electron transitions. These isomers store energy. If these decays could be externally triggered, the energy could be used, but this science is still in its infancy.

OBSOLETE USES OF BROMINE

Leaded gasoline was introduced in the 1930s to help gasoline engines run smoothly. Ethylene dibromide ($C_2H_2Br_2$) was added to the gasoline to scavenge excess lead by the conversion of lead into lead tetrabromide ($PbBr_4$) which was then expelled in the exhaust gases. Unfortunately, that practice introduced substantial amounts of lead into the atmosphere; lead is a toxic element and breathing it is unhealthy.

The 1972 passage of the Clean Air Act by the United States Congress resulted in the introduction of the catalytic converter in automobile exhaust systems to reduce the emissions of carbon monoxide (CO), oxides of nitrogen (principally NO and NO_2—together called "NO_x," where x = the ratio of oxygen atoms to nitrogen atoms for a particular compound), and unburned hydrocarbon fragments (HC). It was discovered, however, that lead poisoned catalytic converters by forming a sticky coating on the catalytic surface that prevented the conversion of CO, NO_x, and HC into harmless substances. Therefore, unleaded gasoline was introduced to solve the problem with catalytic converters. That decision also had the beneficial effect of solving the problem of

(continues on page 44)

FLUORINE, CHLORINE, BROMINE, AND THE OZONE HOLE

Stratospheric ozone (O_3) is present at altitudes ranging from 6 to 30 kilometers (about 4 to 20 miles) above sea level. At these altitudes, O_3 is formed by interaction of cosmic rays with naturally occurring NO_x in the upper atmosphere. Ozone is efficient at absorbing UV radiation and reemitting it in the infrared wavelengths—as heat—which is why the stratosphere's temperature increases with height. Ozone's emission and absorption explain its ability to protect humans, animals, and plants from the sun's harmful ultraviolet wavelengths. Since radiation wavelengths around 300 nanometers have been shown to present a high risk of skin cancer, especially to people whose skin has a low melanin content, the high-level ozone acts as a shield to protect the health of humans as well as virtually all other organisms at the planet's surface.

This natural shield has been damaged, however, by halogenated hydrocarbons—called chlorofluorocarbons (CFCs)—used popularly as refrigerants and aerosol propellants that came on the market in the 1930s. Chlorine atoms are freed from the CFC compounds by collision and interaction with other molecular species in the atmosphere as well as by photon interaction. Atmospheric ozone is then destroyed by the following reaction:

$$Cl + O_3 \rightarrow ClO + O_2$$

When scientists and policy makers recognized the seriousness of the problem in the mid-1980s, an international collaboration led to the Montreal Protocol on Substances that Deplete the Ozone, commonly referred to as the Montreal Protocol, which stipulated a complete phase-out of CFCs by 2005. The substitution of HCFCs (hydrochlorofluorocarbons) for CFCs has helped, but recent studies of the effects of bromine compounds are complicating the ozone-depletion issue.

(continues)

(continued)

Although the concentration of bromine in the atmosphere is at least 100 times smaller than that of chlorine, interactions with bromine atoms destroy about 60 times more ozone. In the lower stratosphere, the percent loss of ozone due to the presence of bromine has been measured to be approximately equal to the loss due to chlorine.

Unlike fluorine, which is extremely stable in the atmosphere because it is mostly bound in HF, bromine is easily detached from its compounds via interaction with light or photolysis, and is therefore readily available in great numbers to interact with O_3 molecules via the following reaction:

$$Br + O_3 \rightarrow BrO + O_2.$$

A major source of bromine on earth is the agricultural fumigant methyl bromide (CH_3Br), which is used to control unwanted vegetation and insects. Recognition of its risk prompted the Montreal Protocol group to mandate elimination of its production as of January 1, 2005. Still, some regions have been granted a so-called critical exemption until suitable substitutes have

(continued from page 42)

atmospheric lead. In fact, since the introduction of unleaded gasoline in 1975—and the subsequent total phaseout of leaded gasoline—the concentration of lead in the atmosphere has dropped almost to zero. Without the lead in gasoline, there has been no need for the ethylene dibromide that used to be in gasoline.

TECHNOLOGY AND CURRENT USES

Bromine compounds have several important uses. For example, bromine has been used in anti-knocking compounds in motor vehicle

been found. Several U.S. states still operate under this exemption, largely owing to their use of methyl bromide in agriculture and in the storage of dried foods.

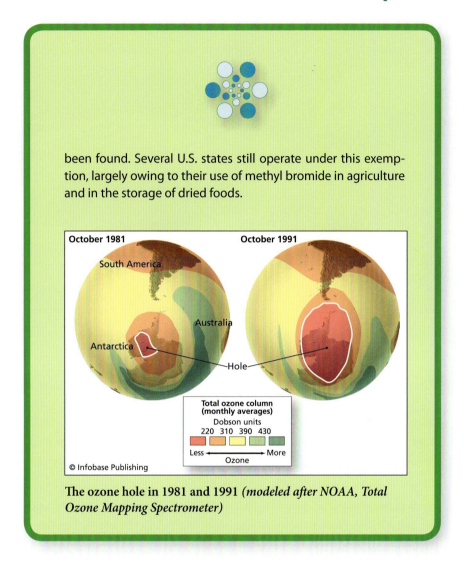

The ozone hole in 1981 and 1991 *(modeled after NOAA, Total Ozone Mapping Spectrometer)*

engines. Bromine is also an oxidizing agent, although one not as powerful as chlorine or fluorine. One class of agricultural fumigants contains methyl bromide and ethylene dibromide, which are used to control unwanted vegetation and insects but present a risk to beneficial ozone. This prompted the Montreal Protocol group to mandate elimination of methyl bromide as of January 2005. Some regions, however, have been granted an exemption until suitable substitutes are found.

Flame retardants may contain chlorine and bromine compounds. Bromine is used in disinfectants and sanitizing agents in swimming pools and drinking water. Because of its bright color, bromine is used

Sodium and potassium bromide are used in development of photographic film. *(iStockphoto)*

in dyes. Bromine compounds are antiepileptic agents used in the treatment of epilepsy. Sodium and potassium bromide are used in film photography; however, with the replacement of film by digital photography, the use of these substances has decreased.

A positron emitter, such as ^{75}Br, injected into a person's bloodstream will eventually reach all areas of the body, including the brain. Positrons are useful in medicine because the brain's activity can be measured by electron current. This is the basis of the PET-scan, which gives a two-dimensional picture of a very particular region or slice of the brain. A PET-scan can look at a multitude of neighboring slices to find possible indications of edema, Parkinson's disease, or lung cancer.

4

Iodine and Astatine: So Alike Yet So Different

Iodine, element 53, is never found in nature as the free element, but there are a few minerals that contain this species. Like other halogens, iodine is an oxidizing agent, although not as powerful as the halogens in the column above it (see the periodic table on page 116.) Iodine is essential to human health in the prevention of goiters. In fact, iodine is the heaviest element essential to health. Artificial isotopes of iodine are used every day in *nuclear medicine,* which is that field of medical practice that uses radioactive materials in the diagnosis and treatment of disease.

In contrast, there are no stable isotopes of astatine (element 85), the heaviest known halogen. What little astatine is present in Earth's crust exists only because minuscule amounts are produced by the radioactive decay of other elements. Because so little is known about astatine,

Iodine crystals are not found in nature, but can be grown in the lab. (*Charles D. Winters/Photo Researchers, Inc.*)

it makes sense to discuss astatine together with iodine rather than to devote a separate chapter to it.

EVERYWHERE AND NOWHERE

Similar in many characteristics, iodine and astatine are, however, not found in anything near equal abundances. Iodine is dispersed throughout the oceans at about 0.05 ppm. Though the process is not yet understood, some seaweed and algae absorb and concentrate iodine in their systems. It is connected to life processes in plants and bacteria, and therefore exists in soil and grasses, and is associated with oil, natural gas, and coal deposits. (Oddly, Australian coal contains on average about three times the iodine concentration as coal from other global sources.)

Astatine, on the other hand, is so rare as to be practically nonexistent in its natural state. Certain isotopes of both elements can evolve from the fission of ^{238}U and ^{235}U, though in iodine's case the latter must be induced by slow neutrons. The difference is that the astatine generally does not last long enough to be noticeable.

THE BASICS OF IODINE

Symbol: I
Atomic number: 53
Atomic mass: 126.90447
Electronic configuration:
  [Kr] $4d^{10}5s^25p^5$

  T_{melt} = 237°F (114°C)
  T_{boil} = 364°F (184°C)

  Abundances:
  In Earth's crust 0.46 ppm
  In seawater 0.05 ppm

Iodine		
2	**I**	113.7°
8	**53**	184.4°
18		546°
18		
7		
	+1+5+7-1	
	126.90447	
	2.9×10^{-9}%	

Isotope	Z	N	Relative Abundance
$^{127}_{53}I$	53	74	100%

THE BASICS OF ASTATINE

Symbol: At
Atomic number: 85
Atomic mass: No stable isotopes.
  Longest-lived isotope has a mass of 210.
Electronic configuration:
  [Xe] $4f^{14}5s^25p^65d^{10}6s^26p^5$

  T_{melt} = 576°F (302°C)
  T_{boil} = unknown

  Abundances:
  In Earth's crust < 1 ppm

Astatine		
2	**At**	302°
8	**85**	
18		
32		
18		
7	[210]	

Isotope	Z	N	$t_{1/2}$
$^{210}_{85}At$	85	125	8.1 hours

The so-called primordial isotope of iodine, ^{129}I, is formed in galactic nucleosynthesis by the astrophysical "r-process," which requires neutron capture at a relatively rapid rate and which generally occurs in supernovae explosions. It has been labeled an "extinct nuclide" because all of the ^{129}I that formed early in the universal timescale has long since decayed to ^{129}Xe. The isotope does exist on Earth, however. It is consistently produced in uranium fission reactions as well as by cosmic-ray interactions in the upper atmosphere. The atmospheric concentration of this radioactive nuclide increased during the era of nuclear weapons testing and continues to be ejected into the environment as a result of the reprocessing of spent nuclear fuel. In this case, the concentration is relatively high. While discharging these radioactive substances into the water table is not considered advantageous, it does have one useful function in that it can be used as a tracer to study water mass movement.

DISCOVERY AND NAMING

Bernard Courtois (1777–1838), born in Dijon, the capital city of Burgundy, France, found his teenage years coinciding with the French Revolution (1789–94). Bernard's father, Jean-Baptiste, was a pharmacist and chemistry lecture demonstrator at the Dijon Academy. Jean-Baptiste worked for Guyton de Morveau, whose chemical research concentrated on finding methods of preparing saltpeter (potassium nitrate, KNO_3) artificially for use in gunpowder. Thus, Bernard grew up in a household well versed in chemistry.

In 1795, Courtois moved to the town of Auxerre, where he apprenticed as a pharmacist and learned additional chemistry. In 1798, he moved to the Polytechnical School in Paris, where he obtained a position in the laboratory of Antoine-François de Fourcroy. In 1799, Courtois was conscripted to serve in the French army, where he served as a pharmacist. He returned to the Polytechnical School in 1801, and in 1802 joined the laboratory staff of Armand Seguin to study the properties of opium.

In 1804, the French emperor Napoléon I Bonaparte issued a decree that militarized the Polytechnical School. Courtois resigned to join his father, who had recently moved to Paris to establish a saltpeter business. When France found itself at war, enough saltpeter from wood ashes could not be obtained to manufacture the quantity of gunpowder (of which saltpeter is a major ingredient) the army required. To make up the difference, saltpeter manufacturers turned to the ashes of seaweed. Unknown to Courtois, seaweed is a rich source of iodine, which at that time had not yet been discovered.

The first indication of iodine's presence in kelp occurred in 1811, when Courtois noticed a violet vapor being formed in the course of his experiments. The copper vessels Courtois used to hold the kelp ashes were being corroded, and Courtois was trying to find an explanation. Courtois had added sulfuric acid with the result that the violet vapor that formed could be condensed into crystals. Courtois did not report his result immediately, but continued to study the properties of this new substance on the side while still pursuing his main business of manufacturing saltpeter. In particular, he discovered that this new substance apparently could not be further decomposed (which would make it an element, not a compound), that it did not react readily with carbon or oxygen, but that it did react with hydrogen and phosphorus. Also, he found that it formed an explosive compound with ammonia.

It was not until 1813 that Courtois's former colleague, Nicolas Clement, reported the discovery, giving Courtois credit. Simultaneously, Humphrey Davy and Joseph-Louis Gay-Lussac (1778–1850) studied the new substance and published papers on its properties. At the time, chlorine was the only other halogen that had been discovered; yet, these chemists quickly concluded that the new substance's properties were very similar to those of chlorine. It was Gay-Lussac who proposed the name *iodine* from the Greek word for "violet."

It was soon discovered that iodine could be used to treat goiters, usually in the form of potassium iodide (KI) or as a mixture of iodine (I_2) and potassium iodide. In 1831, the Royal Academy awarded Bernard Courtois a substantial prize for his discovery of iodine. Unfortunately, Courtois spent the entire sum before he died. Consequently, Courtois—and later his wife—died in poverty.

As the periodic table began to be filled in between bismuth (element 83) and uranium (element 92), chemists recognized that there should be a halogen below iodine. In 1940, a team of scientists at the University of California at Berkeley, led by physicist Emilio Segrè (1905–89),

Emilio Segrè led the team of University of California, Berkeley, scientists that discovered astatine in 1940. *(Lawrence Berkeley National Laboratory/Photo Researchers, Inc.)*

bombarded a sample of bismuth with alpha particles and made a microscopic sample of element 85. Sensitive studies of the new element's chemical properties showed it to be a halogen. Segrè and his coworkers named the new element *astatine,* meaning "unstable," because astatine is the only halogen with no stable isotopes. Astatine's longest-lived isotope, $^{210}_{85}At$, has a half-life of only eight hours.

ON BEING ELECTROPOSITIVE

Like chlorine and bromine, iodine can form both negative ions and chemical bonds in which the iodine is in +1, +3, +5, or +7 oxidation states. The property of possessing the largest atoms of the series F, Cl, Br, and I therefore makes iodine the most electropositive of these halogens. (Being electropositive means that the iodine atoms tend to donate electrons to the covalent bonds rather than being electron acceptors.) Therefore, iodine tends readily to form compounds with more electronegative elements such as fluorine and chlorine, and also with oxygen. Examples include iodine chloride (ICl), iodine bromide (IBr), and various oxyanions such as iodate (IO_3^-) and periodate (IO_4^-). Because of the electropositive nature of iodine, the covalent bonds formed to more-electronegative elements tend to be fairly weak.

THE CHEMISTRY OF IODINE

Since chlorine, bromine, and iodine form a *triad* of elements, iodine's chemistry is very similar to the chemistry of chlorine and bromine, as illustrated by the ions and molecules in the table on page 54.

Like molecular fluorine (F_2), molecular chlorine (Cl_2), and molecular bromine (Br_2), molecular iodine (I_2) is an oxidizing agent. In order of decreasing strength as oxidizing agents, F_2 is more powerful than Cl_2, which is more powerful than Br_2, which in turn is more powerful than I_2. A more powerful molecular halogen can oxidize a halide ion whose molecular form is a less powerful oxidizing agent, as illustrated in the following set of reactions:

$$F_2 + 2\ Cl^- \rightarrow 2\ F^- + Cl_2$$

$$F_2 + 2\ Br^- \rightarrow 2\ F^- + Br_2$$

$$F_2 + 2\ I^- \rightarrow 2\ F^- + I_2$$

$$Cl_2 + 2\ Br^- \rightarrow 2\ Cl^- + Br_2$$

$$Cl_2 + 2\ I^- \rightarrow 2\ Cl^- + I_2$$

$$Br_2 + 2\ I^- \rightarrow 2\ Br^- + I_2.$$

OXIDATION STATES OF THE HALOGENS

	-1	0	+1	+3	+5	+7	-1	-1	-1
chlorine	Cl^-	Cl_2	ClO^-	ClO_2^-	ClO_3^-	ClO_4^-	NaCl	HCl	CCl_4
bromine	Br^-	Br_2	BrO^-	BrO_2^-	BrO_3^-	BrO_4^-	NaBr	HBr	CBr_4
iodine	I^-	I_2	IO^-	IO_2^-	IO_3^-	IO_4^-	NaI	HI	CI_4

As the most electropositive of the halogens, iodine is the weakest oxidizing agent in the series. I_2 will not react with F^-, Cl^-, or Br^-. Iodine is, however, a sufficiently strong disinfectant to kill bacteria without posing the hazards to humans that chlorine and bromine do. Therefore, iodine is the halogen of choice in antiseptics.

ASTATINE CHEMISTRY: WHY THERE IS SO LITTLE

The two most stable isotopes of astatine have half-lives of only 7.2 hours (^{211}At) and 8.3 hours (^{210}At), but these do not occur naturally on Earth. They can be prepared artificially, but only minute quantities can be obtained. Chemical reactions of astatine have been studied. They demonstrate that astatine's chemistry is similar to the chemistry of the other halogens. For example, astatine exists in pure form as diatomic molecules (At_2), and its most stable ion is astatide (At^-).

IODINE AND HYPOTHYROIDISM

Hypothyroidism refers to the physiological condition where the thyroid is deficient in its production of thyroid hormones. The thyroid gland, found in the neck, converts iodine (or, more specifically, iodide) to the hormones tri-iodothyronine (T3) and thyroxine (T4), with T3 being the more biologically active. Insufficient iodine consumption results in a deficiency of T3 and T4 in the body. This can lead to the condition called hypothyroidism, in which some of the body's functions slow down, resulting in symptoms such as lethargy, constipation, depression, and sensitivity to cold temperatures. The thyroid gland may grow larger in an attempt to produce more hormones and lead to the formation

MEDICAL APPLICATIONS OF IODINE ISOTOPES

Iodine 131 was the first radioactive isotope used to diagnose and treat human medical problems, specifically those of the thyroid. Nuclear medicine techniques have now become commonplace as nuclear reactors produce more and more of the needed isotopes.

There are two ways for radiotherapy to be delivered to the location in the body where it will do the most good—by ingestion via normal physiological pathways or by injection of a small source or "seed" to a specific site, a process called *brachytherapy*. The radiation emitted by a radioisotope is like a bullet that, if properly aimed, can damage target receptor cells, and is especially potent against those that divide rapidly like cancer cells.

Iodine, which is preferentially absorbed by the thyroid gland, is a natural candidate in the radiotherapy battle against thyroid cancer. The β-emitter ^{131}I is especially useful, as it has a half-life of only eight days. It can deposit radiation for a short period, inflicting minimal damage to neighboring healthy and useful tissues. This isotope is also used to monitor blood flow through the kidney and to diagnose urinary tract obstruction. Short-term side effects are normally limited to tenderness in the area of the salivary glands and nausea.

Another valuable radioisotope of iodine, ^{125}I, is used in brachytherapy. A small seed of the species can be deposited on or near a malignant tumor to attack and reduce its size. It has the standard procedure in treatment of prostate cancer. Iodine 125, with a half-life of 59 days, emits photons of γ-ray energy (35 keV), but at a low rate, so the cancer is slowly bombarded over a few months. This treatment has made prostate cancer one of the most curable. It may also be useful to treat carcinoma of the pancreas. The easily detectable γ-rays make this isotope indispensable for diagnostic purposes. It can be introduced into the body and its radiation then monitored to test bone density, kidney filtration, and hormone availability.

Another γ-ray-emitting isotope, ^{123}I, which has a half-life of only 13.3 hours, may prove important for other diagnostics. It can only be produced, however, in proton accelerators called *cyclotrons,* making the process rather expensive.

of a goiter. Hypothyroidism is treatable by supplementary T4, which is available in tablet form and which is easily converted by the thyroid to T3. Globally, the condition is less common in North America and Europe, where table salt is iodized, and in Japan, where seaweed is a fair proportion of the diet.

Oddly enough, the opposite cause—a sudden increase in iodine intake—can have the same effect. If a person from a country where little iodine is consumed moves to a country where much of the food consumed contains iodine, the thyroid will initially convert the excess iodine into thyroid hormone, but will soon shut down, so as not to introduce excess hormone into the system. This protective method of regulation does not always have a long-term effect, as the gland can start functioning again in some individuals. In others, however, the shutdown of hormone production leads to hypothyroidism in a more serious form, one in which the supplementation by T4 will probably not be sufficient.

On the other hand, if excess iodine intake does not shut down the thyroid gland, it can lead to hyperthyroidism—the case of overactivity of the thyroid gland, where excess hormone is released. Symptoms can include feelings of anxiety and of being too warm. The condition is treatable by reducing iodine intake. A "hidden" form of sudden iodine intake can occur during medical procedures that use tracer dyes containing iodine, with hyperthyroidism as a possible result.

An extreme, but perhaps unrelated, reaction can result from consumption of shellfish. The iodide ion in seawater concentrates in marine organisms. Iodide can build up to high concentrations in shellfish. Shellfish allergies, sometimes resulting in anaphylactic shock, have sometimes been equated to iodine allergies, but medical researchers remain undecided on this important cause/effect relationship.

ASTATINE'S RADIOACTIVE NATURE

Astatine is unique in that it is the only halogen with no stable isotope. Each of its 33 known isotopes decays fairly quickly. The naturally occurring astatine isotopes produced by the radioactive decay of ^{238}U and ^{235}U are radioactive themselves, with half-lives ranging from nanoseconds to seconds, and subsequently decay to lead. So even though astatine is continuously coming into existence, it generally does not last long enough to be noticeable.

As such, astatine has by and large been considered to be unimportant. Early references state that it has no useful properties. The alpha-decay properties of astatine do, however, seem to have useful applications.

A couple of fairly long-lived isotopes that do not occur naturally, ^{210}At ($t_{1/2}$ = 8.1 hours) and ^{211}At ($t_{1/2}$ = 7.2 hours), can be produced in the laboratory by bombarding bismuth atoms with alpha particles. Astatine 211 is useful in some types of cancer treatment. The high-energy α-particles it emits can be made to bombard a localized tumor, and its short half-life ensures that it will not remain in the body for an extended period of time. Its daughter particles also decay rapidly, so the risks to the patient are minimal. Unfortunately, it is difficult to produce pure ^{211}At; there is always some fraction of ^{210}At also present. Astatine 210 decays to the highly radioactive polonium 210, a strong emitter of α-particles that has a half-life of 138 days. It is highly undesirable to have ^{210}Po in the system, so researchers continue to work on a means of purifying ^{211}At samples for medical use.

TECHNOLOGY AND CURRENT USES

Iodine is an important *bactericidal,* a substance that kills harmful bacteria. It can be used to sterilize wounds and drinking water. Radioactive iodine can be used to treat thyroid conditions.

Iodine can also be used as an X-ray contrast medium, where a dye injected into a blood vessel—an artery or vein—allows X-ray pictures to be taken of the blood vessel.

Silver iodide is an important component of photographic film. Dark images form on the film when light strikes the film and causes the silver ions to be reduced to silver metal. Silver iodide is also used in cloud seeding, where it constitutes particles on which raindrops can form.

Iodine's absorption and refraction of light is also useful. It can color foods red or brown, and astronomers use iodine gas absorption cells to calibrate stellar spectra.

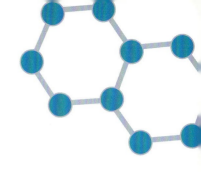

PART

II

The Noble Gases

INTRODUCTION TO THE NOBLE GASES

In 1869, when Dmitri Mendeleev developed his periodic table, the noble gases had not yet been discovered. In fact, their existence was not even suspected. Because the concept of atomic number had not yet been discovered, there did not appear to be any "gaps" in Mendeleev's table where the noble gases are now located. It appeared very reasonable that sodium followed fluorine, potassium followed chlorine, rubidium followed bromine, and so forth; the family of halogens was followed by the family of alkali metals in a very orderly fashion.

During the 1800s, the noble gases—helium (He), neon (Ne), argon (Ar), krypton (Kr), xenon (Xe), and radon (Rn)—were discovered. Scientists found that all of the noble gases exist on Earth as products of radioactive decay of other elements. All of the noble gases are components of the atmosphere, although radon isotopes are very short-lived, so there is almost no radon in the atmosphere. The noble gases all have normal melting and boiling points well below room temperature and are liquids over a range of only a few degrees. Because they do not exist in the atmosphere in chemical combination with any other elements, it was natural to believe that they are chemically *inert,* or unreactive. In fact, the noble gases exist in the atmosphere only as *monatomic* gases; that is, a noble gas molecule consists of a single atom rather than two atoms like N_2 and O_2. In addition, for many years, no one could find any examples of a noble gas forming compounds with other elements—either in nature or in the laboratory. The chemical inertness of the noble gases is attributed to their electronic configurations. The lightest noble gas, helium, has a configuration of $1s^2$, which means its only electrons are in a filled energy level, and helium does not form chemical bonds. The remaining noble gases have configurations that end in the pattern ns^2np^6, where n is the energy level. These eight electrons form an energetically stable "octet" such that the atoms have little or no tendency to form chemical bonds. They tend neither to donate electrons to other atoms nor to gain electrons from other atoms; neither do they readily share electrons covalently with other atoms.

That picture changed, however, in 1962, when Neil Bartlett, a Canadian chemist at the University of British Columbia, quite elegantly made XeF_2 by combining xenon gas with fluorine gas in the presence of sunlight. Later, Bartlett moved to the University of California at Berkeley. He, along with scientists around the world, subsequently succeeded in making other compounds of xenon and krypton—mostly fluorides, oxides, and oxyfluorides. To date, however, no compounds have been made with helium, neon, or argon.

DISCOVERY AND NAMING OF THE NOBLE GASES

When Dmitri Mendeleev published his periodic table, the table was one column short of modern versions of the table. In Mendeleev's table, the last column would have been the halogens. The reason the noble gases had not yet been discovered is due to their relative chemical inertness. They are not found in any compounds in the ground, in the seas, or in the atmosphere. Nor do they form any compounds with oxygen, acids or bases, or other reagents that were in use in the 19th century. Even after the isolation of fluorine gas, 76 years elapsed before anyone thought to try reacting fluorine with any of the noble gases. Thus, for the first 70 or so years that the noble gases were known, they were called the *inert gases* because it was believed that they would not form any chemical compounds at all. Ironically, in the 1930s, Enrico Fermi's group in Italy missed discovering fission because they failed to recognize that one of the common products of fission is the noble gas krypton, which they did not suspect would be formed in their experiments and which escaped from the reaction vessel undetected because of its inertness.

Helium, the lightest noble gas, holds the distinction of being the only element to have been detected completely outside Earth's environment before being discovered here on Earth. In 1868, astronomers Pierre-Jules-César Janssen (1824–1907) of France and Sir Joseph Norman Lockyer (1836–1920) of England independently observed spectral lines in the Sun's spectrum that could not be attributed to any element known on Earth at that time. (Janssen made his discovery while in India observing a total eclipse of the Sun.) Lockyer proposed calling the new element helium, from the Greek word *helios* meaning "Sun." At first, scientists did not believe in the existence of helium. It was not until 1895, when Scottish chemist Sir William Ramsay (1852–1916) discovered helium in minerals containing uranium, that Janssen and Lockyer's discovery was accepted.

In 1785, Henry Cavendish had succeeded in absorbing all of the nitrogen and oxygen gases from the atmosphere, but had recognized that a small amount of gas remained that he could not react with anything. Cavendish estimated that this unknown gas had to constitute less than 1 percent of the atmosphere, but he was unable to identify its nature. It was more than a century later before this unknown gas was identified, given the name *argon,* and shown to constitute 1 percent of the atmosphere. The discovery began with experiments performed by John William Strutt (1842–1919), the third Lord Rayleigh, who detected an anomaly in the densities of nitrogen gas prepared by two different methods. The first method was the preparation of nitrogen by the decomposition of ammonia. The second method was the preparation of nitrogen by the removal of oxygen, carbon dioxide, and water vapor from air. The pure nitrogen obtained from ammonia was slightly denser than the nitrogen obtained from air, because the sample from air also contained argon, which is lighter than nitrogen. Rayleigh could not explain his results. In fact, he suspected it might be a different form of nitrogen, N_3, in analogy to ozone, which is O_3.

Rayleigh then began working with Ramsay. Together they showed that the unknown gas had spectral lines that did not match the lines of any known element. Therefore, it had to be a new element. They also demonstrated its chemical inertness. It was Ramsay who proposed that there might be a column of gases that would fit into the periodic table next to the halogens. In 1894, Rayleigh and Ramsay jointly presented their findings to the British Association. It was the chairman of the meeting, H. D. Madan, who suggested naming the new element "argon," meaning "the lazy one."

The following year, 1895, saw Ramsay's discovery of helium in a mineral of uranium called clevite. He found that an inert gas was evolved when clevite was dissolved in acid. Ramsay collected some of the gas in a tube and, along with Lockyer and Sir William Crookes, observed that the gas's spectral lines matched those of helium observed in the solar spectrum. In the same year, the German scientist H. Kayser discovered helium in the atmosphere. Also, two Swedish chemists, Nils Adolf Erik Nordenskiöld (1832–1901) and his student, Nils Abraham Langlet, concurrently obtained helium from clevite, but after Ramsay had already announced his discovery.

Rayleigh, Ramsay, and Ramsay's assistant, Morris William Travers (1872–1961), continued to search for additional inert gases. In 1898, they discovered krypton by diffusing a sample of argon in the hope of separating it into different gases. Krypton was found to have bright yellow and green spectral lines. The name *krypton* was chosen because it means "hidden," and certainly it was hidden in the argon sample. After placing helium, argon, and krypton into a new column of the periodic table, the scientists recognized that there should exist another inert gas that would be located between helium and argon in the table. Thus, also in 1898, using liquefied air, they solidified argon. As the argon volatilized, they collected the first gas that appeared and showed that its bright crimson line was unique. Ramsay's 13-year-old son suggested calling the new gas novum, a name that Ramsay liked, although he changed the name to neon, which also means "new gas."

Finally, also in 1898, Ramsay and Travers separated a still-heavier gas from krypton that had a beautiful blue spectral line. They named this element xenon, meaning "the stranger." For their work, in 1904, Rayleigh was awarded the Nobel Prize in physics, while Ramsay was awarded the Nobel Prize in chemistry.

Radon was discovered by a completely independent path of investigation—the study of radioactivity. Radioactivity had only been discovered in 1896. In 1900, it was still a poorly understood phenomenon. However, in that year, Friedrich Ernst Dorn showed that one of the disintegration products of radium is an inert gas that at first was called radium emanation, or niton. Later, the name was changed to radon. In 1910, Ramsay measured the density of radon and showed that it is the heaviest gas known.

In August 2006, the discovery of element 118 was reported. Element 118 lies in the column of noble gases in the periodic table. However, because the boiling points of elements increase as one descends a column of the table, the boiling point of element 118 at one atmosphere pressure should be above room temperature. A sample of element 118 large enough to measure its boiling point has not been prepared at the time of publication of this volume, but it is predicted that it will actually be a noble liquid, not a noble gas.

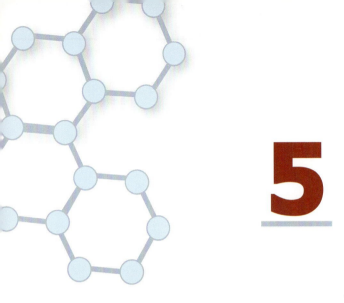

5

Helium: From Balloons to Lasers

Helium—element number 2—is the second most abundant element in the universe (after hydrogen). Hydrogen atoms do not require the presence of a neutron for stability. However, because helium nuclei have two protons—and protons have a strong electrostatic repulsion—a helium nucleus must have at least one neutron. An attraction between neutrons and protons, called the *strong nuclear force,* overcomes the force of repulsion between the protons and holds the helium nuclei together. The same requirement is true of all the remaining elements in the periodic table; their nuclei must have at least one neutron present.

A remarkable fact about helium, as well as neon and argon, is that no chemical compounds have ever been made with it. Helium can be ionized in the gas phase to the He^+ and He^{2+} ions, but helium atoms

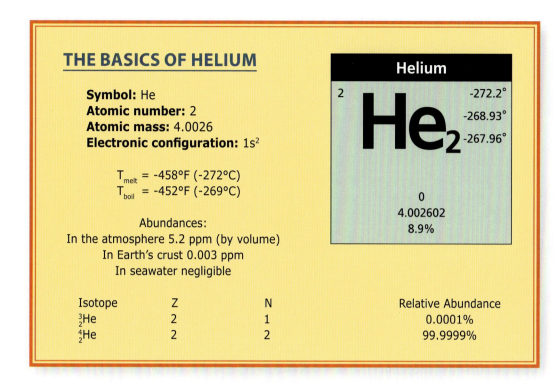

THE BASICS OF HELIUM

Symbol: He
Atomic number: 2
Atomic mass: 4.0026
Electronic configuration: $1s^2$

T_{melt} = -458°F (-272°C)
T_{boil} = -452°F (-269°C)

Abundances:
In the atmosphere 5.2 ppm (by volume)
In Earth's crust 0.003 ppm
In seawater negligible

Isotope	Z	N	Relative Abundance
$^{3}_{2}$He	2	1	0.0001%
$^{4}_{2}$He	2	2	99.9999%

Helium

2

He$_2$

-272.2°
-268.93°
-267.96°

0
4.002602
8.9%

do not form chemical bonds to other helium atoms or to atoms of any other element. Consequently, there is no "chemistry" of helium. Any study of helium involves the study of its physical properties, with no chemical properties to speak of.

Since hydrogen and helium are the two lightest gases, they have been the choices for balloons and airships. Hydrogen, however, is flammable, and its use resulted in several explosions, the most famous being the destruction of the *Hindenburg* in 1937. Even after the discovery of helium in the atmosphere in the 1890s, insufficient helium was available for commercial use. It was not until the 1930s that helium could be recovered economically from underground deposits. From then on, helium replaced hydrogen in dirigibles.

Today, with the notable exception of the Goodyear blimp, commonly seen hovering over football stadiums, dirigibles tend to be a thing of the past. There are more important uses of helium, however, than just to fill balloons. Much more important are the *cryogenic* uses of helium to keep laboratory instrumentation in laboratories and hospitals at temperatures very near absolute zero.

THE ASTROPHYSICS OF HELIUM

In the Sun and all other stars of similar mass and age, extreme heat and pressure allow the continual fusion of hydrogen to helium in the stellar core. The process is called the *proton-proton chain* (PP chain). Fast-moving hydrogen nuclei (protons), packed closely together, collide millions of times per second. Some combine to form deuterium, which can then collide with a proton to make ^{3}He or *light helium,* which can then collide with one like itself to finally make ^{4}He. The chain reaction is written as follows:

$$_1^1p + _1^1p \rightarrow _1^2H + e^+ + \nu_e + 0.42 \text{ MeV}$$

$$_1^2H + _1^1p \rightarrow _2^3He + \gamma + 5.49 \text{ MeV}$$

$$_2^3He + _2^3He \rightarrow _2^4He + 2\,_1^1p + 12.86 \text{ MeV},$$

where p symbolizes proton, e$^+$ positron, γ a photon, and ν_e an *electron neutrino.*

The energy in MeV is the kinetic energy of motion the product particles gain, and the γ is a *photon* of zero mass, but measurable energy that depends on its frequency. Energy appears at the expense of the mass that is lost in the sequence according to Einstein's equation $E = mc^2$. Both the kinetic and photon energy can help to regulate the temperature at the stellar core. All fusion reactions of elements lighter than iron give off energy in one form or another. (Except where needed to make a specific point, kinetic energies are omitted in the following discussion.)

The PP chain is the most effective producer of helium in the Sun, but a couple of other methods help. The PPII chain (sequence follows) contributes about a third of the helium in the Sun:

$$_2^3He + _2^4He \rightarrow _4^7Be + \gamma$$

$$\mathrm{^{7}_{4}Be + e^{-} \rightarrow {}^{7}_{3}Li + \nu_{e}}$$

$$\mathrm{^{7}_{3}Li + {}^{1}_{1}H \rightarrow {}^{4}_{2}He + {}^{4}_{2}He.}$$

A third process, the following PPIII chain, is rare in the Sun, producing only 0.1 percent of its helium, but is dominant in more massive stars where the stellar core temperature exceeds 23 million Kelvin, which allows beryllium to fuse with hydrogen more readily:

$$\mathrm{^{7}_{4}Be + {}^{1}_{1}H \rightarrow {}^{8}_{5}B + \gamma}$$

$$\mathrm{^{8}_{5}B \rightarrow {}^{8}_{4}Be + e^{-} + \nu_{e}}$$

$$\mathrm{^{8}_{4}Be \rightarrow 2\, {}^{4}_{2}He.}$$

Helium came into being long before stars formed, however. The entire universe was hot and dense enough just after the big bang for helium to be formed in several other ways. Free-flying neutrons (n), protons (p), and deuterons (d) colliding in great numbers could fuse with one another to make ^{3}He and ^{4}He in the following reactions:

$$\mathrm{^{2}_{1}d + {}^{2}_{1}d \rightarrow {}^{3}_{2}He + {}^{1}_{1}n} \quad \text{followed by} \quad \mathrm{^{2}_{1}d + {}^{3}_{2}He \rightarrow {}^{4}_{2}He + {}^{1}_{1}p}$$

or

$$\mathrm{^{2}_{1}d + {}^{2}_{1}d \rightarrow {}^{3}_{1}H + {}^{1}_{1}p} \quad \text{followed by} \quad \mathrm{^{2}_{1}d + {}^{3}_{1}H \rightarrow {}^{4}_{2}He + {}^{1}_{0}n.}$$

There were also slower processes involving photon emission:

$$\mathrm{^{2}_{1}d + {}^{1}_{0}n \rightarrow {}^{3}_{1}H + \gamma} \quad \text{followed by} \quad \mathrm{^{3}_{1}H + {}^{1}_{1}p \rightarrow {}^{4}_{2}He + \gamma}$$

or

$$\mathrm{^{2}_{1}d + {}^{1}_{1}p \rightarrow {}^{3}_{2}He + \gamma} \quad \text{followed by} \quad \mathrm{^{3}_{2}He + {}^{1}_{0}n \rightarrow {}^{4}_{2}He + \gamma.}$$

Such a wide variety of formation methods made helium the second most abundant element in the universe. Nevertheless, it is not

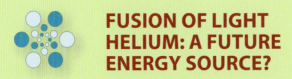

FUSION OF LIGHT HELIUM: A FUTURE ENERGY SOURCE?

While deuterium-tritium fusion has been considered the most likely process to result in a fusion reactor suitable for electricity production, and much research has gone into the attempt, some scientists and entrepreneurs are now focusing attention on the fusion of deuterium with light helium (^{3}He)

$$_1^2d + {}_2^3He \rightarrow {}_1^4He + {}_1^1p + 18.4 \text{ MeV}.$$

It is an attractive notion in that only a tiny fraction of the by-product would be fast neutrons. (These would result from d-d and p-p collisions.) So a light-helium fusion reactor would be much cleaner in terms of harmful radiation. It is also more feasible for electricity production because both energy-carrying products are charged particles.

The problem is that Earth has an extreme paucity of light helium. Of all the helium in all deposits, only one ten-thousandth portion is ^{3}He. While a tiny portion of this isotope comes from the decay of tritium in thermonuclear weapon stockpiles, overall the dearth is so overwhelming as to make the idea of using it seem almost nonsensical.

To understand the issue, it is useful to ask why there is so little helium on earth. The sun continuously produces it and expels large portions of its atmosphere on the solar wind, so Earth should be receiving an unending supply. Where is it? The answer is that the helium nuclei delivered by the solar wind are charged particles that are either deviated by Earth's magnetic field or intercepted by particles in the upper atmosphere, so none reaches the surface.

easy to find on Earth. Terrestrial helium forms only in natural gas deposits that evolved in the proximity of radioactive thorium and uranium ores, and there are few places on Earth that it occurs in sufficient quantity to make collection worthwhile. What helium is not

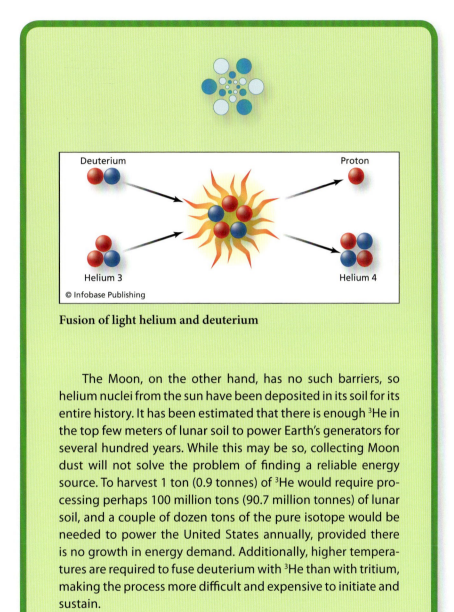

Deuterium   Proton

Helium 3   Helium 4

© Infobase Publishing

Fusion of light helium and deuterium

The Moon, on the other hand, has no such barriers, so helium nuclei from the sun have been deposited in its soil for its entire history. It has been estimated that there is enough ^{3}He in the top few meters of lunar soil to power Earth's generators for several hundred years. While this may be so, collecting Moon dust will not solve the problem of finding a reliable energy source. To harvest 1 ton (0.9 tonnes) of ^{3}He would require processing perhaps 100 million tons (90.7 million tonnes) of lunar soil, and a couple of dozen tons of the pure isotope would be needed to power the United States annually, provided there is no growth in energy demand. Additionally, higher temperatures are required to fuse deuterium with ^{3}He than with tritium, making the process more difficult and expensive to initiate and sustain.

purposely extracted during the refining of such natural gas is lost to the atmosphere. Like oil and natural gas, helium is a nonrenewable commodity that some experts predict the world will use up within 25 years. Such an eventuality would pose a serious problem for a host of

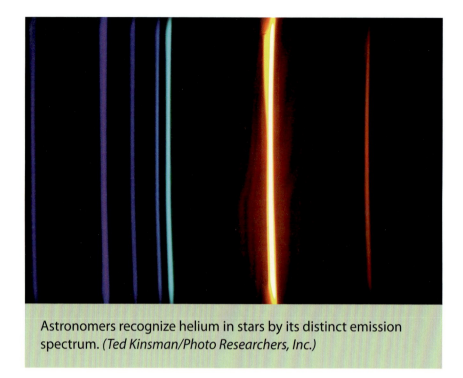

Astronomers recognize helium in stars by its distinct emission spectrum. *(Ted Kinsman/Photo Researchers, Inc.)*

industries and scientific researchers who require helium's important cooling properties.

EARTHBOUND IN NATURAL GAS

Helium exists as two stable isotopes: helium 3 and helium 4. Almost all of the helium on Earth is helium 4, produced by the decay of radioactive elements found in the crust of the Earth. The principal radioactive elements are uranium and thorium, which have such long half-lives that primordial amounts of the two elements still remain from the time of Earth's formation. All of their isotopes are radioactive. Some of their isotopes decay by emitting alpha particles, which are nuclei of helium 4 and which are absorbed into the surrounding rock substrate. The daughters of the radioactive decays are radioactive, and some of them also emit alpha particles.

Over geologic time, the amount of helium in Earth's crust has accumulated, but it is usually found in usable quantities only in association with natural gas deposits. In the United States, the largest is near Amarillo, Texas. With the invention and production of the auto-

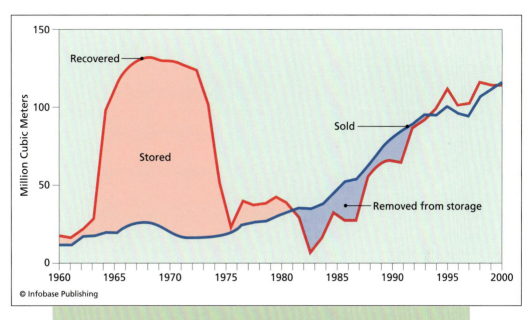

Helium recovery in the United States from 1960 to 2000

mobile early in the 20th century, oil wells were drilled throughout the southern plains states. In the process, huge deposits of natural gas were discovered. In addition, it became apparent that substantial amounts of helium gas were also present. Previously, the only source of helium had been the liquefaction of air. However, because the amount of helium in the atmosphere is minuscule, helium obtained in this manner is extremely expensive. Only when large amounts of helium became available from underground reservoirs did the price of helium drop so that it became feasible to use in commercial applications.

LIGHTER THAN AIR: HELIUM BALLOONS AND AEROSTATICS

In the 1780s, the technology to construct fabrics suitable for balloons and to fill those balloons with hot air that was lighter than the surrounding air initiated the practice of hot-air ballooning, a practice that continues to this day, at least for recreational purposes. At the same time, the discovery of hydrogen gas and the lifting force exerted

Toy balloons filled with helium *(Steven Kratochwill/Shutterstock)*

because of hydrogen's light weight initiated a 150-year period of hydrogen-filled balloons and dirigibles, or "airships." The basic principle behind the physics of lighter-than-air flight is that filling a balloon or other airship frame with a gas—like hydrogen or helium—that is much less dense than air creates a force of *buoyancy,* or lift, so that the airship "floats" in air much in the same way that an object that is lighter than the water it displaces experiences a force of buoyancy that causes it to float.

With the discovery in the first quarter of the 20th century of large underground reservoirs of helium in the United States—and the development of the technology to extract that helium without considerable losses—helium replaced hydrogen as the gas of choice in airships. As evidenced by the explosion of the hydrogen in the airship *Hindenburg,* hydrogen was dangerous. It was much safer to substitute helium, a nonflammable gas, with little sacrifice of lifting force.

HOW HELIUM CHANGES VOICE SOUND

Vocal chords in humans and animals are frequency generators. The frequency depends on tension in the muscles in the throat as well as the shape and depth of the folds of tissue in the vocal chords. The particular tone and pitch are determined by the shape of the cavity in which the sound develops and resonance is formed: The shape can be changed by moving the jaw and tongue, which allows for different noises like words and roars.

The medium for the sound is normally air, which is mostly nitrogen. In air at a standard temperature of 68°F (20°C), a sound wave travels at about 760 MPH (340 m/s). Helium is much lighter than air, however. When the vocal chords thump on the atoms or molecules, a much higher velocity is imparted to the lighter species. (Imagine the difference between thumping a baseball or a ping-pong ball. The ping-pong ball flies off much faster because momentum is conserved in the collision.) Hence, the velocity of

(continues)

A common activity at parties is for someone to inhale helium from a party balloon, which results in the person talking with a high-pitched voice. *(Pool/Reuters/Corbis)*

(continued)

sound in helium, at 2,074 MPH (927 m/s), is more than two and a half times that in air. The wavefronts consisting of helium atoms arrive at the cavity exit—the lips, in humans—much faster than wavefronts consisting of air molecules.

When a listener hears the wave that was generated in helium, the effect is similar to *Doppler shift* of a frequency. Doppler shift happens when an object emitting a sound is moving toward or away from the listener. In the case of approach, each wavefront is emitted closer to the preceding wavefront than it would have been if the object were stationary. So, to the listener, it seems like the frequency is higher. If the object is moving away, it's the opposite; the frequency sounds lower. The classic example is that of a train blowing its whistle as it passes a station. As it approaches, the whistle seems to become higher in pitch; as it moves away, it sounds lower, even though the source is emitting at a constant frequency.

Because a sound wave produced in helium has wavefronts closer together than an equivalently produced vibration in air, the listener perceives the frequency as higher because it is emitted that way from the lips, even though the wave travels through air to the ear.

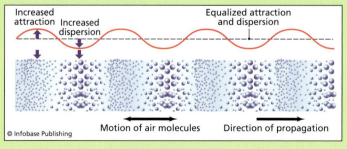

Schematic of air molecules in a sound wave

In 1924, a transcontinental test flight was made of an airship filled with helium. The U.S.S. *Shenandoah* flew from New Jersey to California to Washington and back to New Jersey. At that time, however, helium was so scarce and expensive that there was only enough helium to fill a single airship. On a subsequent flight in 1925, the *Shenandoah* encountered a violent storm and broke apart in midair, and the various parts—with airmen inside them—fell to the ground. Because the ship was filled with helium, instead of hydrogen, there were no explosions. Most of the crew, 29 of 43 airmen, managed to survive the ordeal.

Because of the military situation that developed in Europe in the early 1930s, the export of helium from the United States was banned, preventing Europeans—principally the Germans—from using helium in their zeppelin airships. Despite the ban, until the outbreak of World War II, the Germans continued to fly zeppelins (hydrogen filled) regularly between Frankfurt and New Jersey and to Brazil. For the most part, the Germans had a very fine safety record, with the most notorious exception being that of the loss of the *Hindenburg* in 1937.

In the post–World War II years, passenger service by airships came to an end. Their popularity declined as the popularity of airplane travel in the postwar years increased. For a few years, the U.S. Navy continued to fly airships on reconnaissance missions, but a crash in 1960 ended that program. The descendants of airships today—all filled with helium—are the blimps sometimes used for advertising purposes or for aerial photography.

TECHNOLOGY AND CURRENT USES

Because it is lighter than air, helium gas is used in weather balloons, party balloons, and in high-flying balloons used to take scientific measurements. Some scientists and entrepreneurs are now focusing attention on the fusion of deuterium with light helium as a candidate for fusion power generation. A light helium fusion reactor would be much cleaner in terms of harmful radiation. It is also more feasible for electricity production because both energy-carrying products—a proton and an alpha particle—are charged particles.

To dilute oxygen delivered to scuba divers, helium gas is used in place of nitrogen, which is associated with the bends and nitrogen narcosis. Liquid helium is also important for cooling purposes in space-based experiments and Earth-based facilities—especially for cooling laboratory instrumentation in laboratories and hospitals.

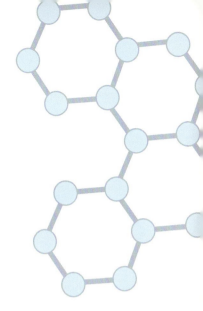

6

Neon: Known for Its Glow

Neon, element number 10, is the second most abundant noble gas in Earth's atmosphere. Like helium, neon has no chemistry. However, neon's emission spectrum is well known for its intensely bright lines in the red and orange regions of the visible spectrum, hence its use in illuminated outdoor signs.

THE ASTROPHYSICS OF NEON

Neon is synthesized in any star with a mass greater than about four times that of the Sun. When all the helium in the core of such a star has fused to carbon, the core will collapse once more until the temperature, raised by gravitationally *infalling* matter, reaches 6×10^8 K—hot enough to fuse carbon via the following reaction:

$$^{12}_{6}C + {}^{12}_{6}C \rightarrow {}^{20}_{10}Ne + {}^{4}_{2}He.$$

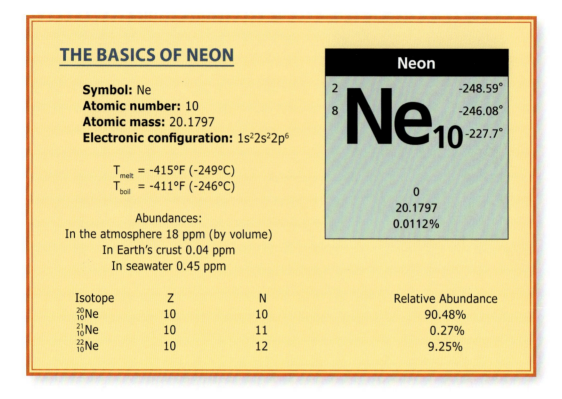

THE BASICS OF NEON

Symbol: Ne
Atomic number: 10
Atomic mass: 20.1797
Electronic configuration: $1s^2 2s^2 2p^6$

T_{melt} = -415°F (-249°C)
T_{boil} = -411°F (-246°C)

Abundances:
In the atmosphere 18 ppm (by volume)
In Earth's crust 0.04 ppm
In seawater 0.45 ppm

Isotope	Z	N	Relative Abundance
$^{20}_{10}$Ne	10	10	90.48%
$^{21}_{10}$Ne	10	11	0.27%
$^{22}_{10}$Ne	10	12	9.25%

Neon

2
8

Ne$_{10}$

-248.59°
-246.08°
-227.7°

0
20.1797
0.0112%

Almost all the neon thus produced is itself subsequently fused into heavier atomic nuclei within the star, however, so any neon observed nowadays must have been generated from a different source.

The sources of current neon in the universe are sporadic, but reliable. Sooner or later, most stars will fuse and collapse until they reach the supernova stage and explode. Countless stars have undergone this transition in the past, and countless more will do so in the future. In the heat and chaos of each supernova explosion, many different particles and nuclei collide to form the myriad heavy atomic species. Neon's formation in supernovae is an alpha process resulting from the collision of oxygen with *alpha particles,* as follows:

$$^{16}_{8}O + {}^{4}_{2}He \rightarrow {}^{20}_{10}Ne + \gamma.$$

While most astrophysicists concur that this reaction is the source of neon in the interstellar medium and thus in stars that formed fairly recently, there is controversy over its abundance in the Sun. Neon is believed to play an important part in solar *convection,* and knowing its

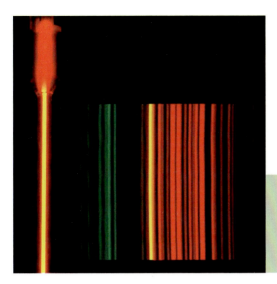

The neon emission spectrum displays a large number of red lines. *(Department of physics and astronomy, Georgia State University)*

abundance in the Sun helps predict the extent of the convection zone. Calculations of the behavior of the convection zones, however, do not agree with the formerly accepted value of neon abundance, which is quite difficult to measure experimentally. Some scientists have urged comparison with more easily obtainable data from stars that seem Sun-like, while others want to stick to strict solar data. It will be an interesting arena for future research.

Other important areas of exploration in astrophysical neon research include its content in meteorites and in preplanetary solar system disks.

EARTHBOUND: ISOTOPES IN VOLCANOES

When the Earth was formed, a portion of the same elements that make up the Sun accreted to make the Earth and other planets in our solar system. As is the case with all noble gases, however, an additional source of neon on Earth is radioactive decay. The most abundant radioactive materials in Earth's interior are ores of uranium and thorium, both of which are emitters of alpha particles. For example:

$$_{92}^{238}U \rightarrow {}_{90}^{234}Th + {}_{2}^{4}He,$$

where $_{2}^{4}He$ is an alpha particle. Alpha particles are emitted from uranium atoms with high kinetic energies, making it possible for them to

BRIGHT CITY LIGHTS

Applying a sufficient electrical voltage difference between the ends of a tube containing any gas will cause that gas to glow because the energy provided by the voltage difference causes the gas to ionize. Electrons are removed from the atoms of the gas, leaving behind positively charged ions. This gaseous mixture of positive ions and negative electrons provides the conducting medium necessary to allow electrical current to flow between the ends of the tube.

The electrons constantly recombine with the ions to form neutral atoms, while just as rapidly the atoms are being ionized again. However, just as electrons have to gain energy to be removed from atoms, the principle of conservation of energy requires that energy has to be emitted for the electrons to recombine with the ions again. The energy the atoms emit is in the form of photons of light, which can be in the infrared, visible, or ultraviolet regions of the spectrum, depending on how energetic the photons are. Of course, the human eye can only perceive visible radiation; any infrared or ultraviolet rays are invisible to the human eye.

Various gases could be used in lighting fixtures, and there are other elements, such as mercury or mixtures of the other noble gases, that are used. Neon, however, is often preferred for outdoor, nighttime lighting—especially in advertising—because the light emitted by neon is a brilliant red color that readily attracts attention. The human eye is especially sensitive to color in that part of the spectrum, making neon lights real attention getters.

The invention of neon lamps is attributed to Georges Claude (1870–1960), a French inventor who unveiled the first neon lamp

bombard nearby nuclei of other elements, combine with those nuclei, and form nuclei of new elements. (Alternatively, the helium ion can acquire two electrons to become a neutral helium atom. This is the source of the deposits of helium found underground in Texas.)

in Paris on December 11, 1910. Five years later, Claude received a patent from the U.S. Patent Office. By the 1920s, Claude's company was producing neon lamps for companies in the United States. The first two lamps used in advertising were purchased by a Packard car dealership in Los Angeles, California, for $24,000—a sum equivalent in 2007 to $295,000 when adjusted for inflation.

Neon gas provides the red color in these Las Vegas lights. (SoleilC/Shutterstock)

To form neon, an alpha particle must strike a nucleus of oxygen. The following reaction results in the formation of neon 21:

$$_2^4\text{He} + {}_8^{18}\text{O} \rightarrow {}_{10}^{21}\text{Ne} + {}_0^1\text{n},$$

where n is a neutron. Alternatively, a neutron could strike a magnesium atom, as in the following reaction:

$$_0^1n + {}_{12}^{24}Mg \rightarrow {}_{10}^{21}Ne + {}_2^4He.$$

Although most of the primordial neon accreted from the solar disk would have been neon 20 and neon 22, the relative amount of neon 21 is gradually increasing from radioactive decay. Through various geological processes that act over very long periods of time, rocks in the Earth's mantle melt and flow to the surface (as in volcanoes). Dissolved gases, including neon, *outgas* from this molten material and escape into the atmosphere, slowly changing its composition.

THE HELIUM-NEON LASER

The first gas laser, invented at Bell Labs in 1960, was a helium-neon laser. This type of laser is still in use today, despite the greater efficiency and enhanced power of other more recent types, largely because it is inexpensive and has a strong, narrow, far-reaching beam. The term *laser* was coined by taking the first letter from each relevant word in the process of making the beam: **L**ight **A**mplification by **S**timulated **E**mission of **R**adiation.

The radiation that is stimulated and amplified is a result of de-excitation of neon atoms. How they become excited is key. In order to have a continuous emission of bright light, an endless supply of atoms excited to the proper energy level is required. Most excited states of atoms do not last long; the electrons have an affinity for a lower energy state, and the atom de-excites rapidly to either the ground state or some state with a lower energy. There are, however, somewhat anomalous configurations, called metastable states, which can remain excited for quite long periods on the atomic time scale (contrast milliseconds with nanoseconds). This comparatively stable atomic configuration comes about because of changes to electron spin orientation. The atom cannot stay in the metastable state forever, though, and eventually the electrons jump down to a lower-energy state, giving off light in the process.

In the helium-neon laser cavity, this state is induced in the neon by a clever application of frequency matching. The tube contains not only both gases, but also a beam of electrons that is produced by simply placing oppositely charged electrodes at each end. The *cathode* is a metal plate heated so that electrons attached to the metal atoms at the surface acquire enough energy to escape and then accelerate toward the *anode*. They collide with helium and neon atoms as they go. The electrons can excite many different states of the atoms, but the potential difference between the electrodes and density of atoms is set so that the electrons preferentially excite helium from its ground state to a state that very nearly matches that of the important metastable state of neon. The excited helium atoms colliding with neon atoms transfer that excitation so that a *population inversion* is achieved, whereby there are more neon atoms in the metastable state than in a lower energy state. So when the excited atom decays by emitting a photon, there are very few lower-energy neon atoms to absorb it.

Instead, the ejected photon, upon encountering a neon atom in the metastable state (of which there is a large population), can stimulate its decay, forcing the neon atom to eject a photon of exactly the same wavelength and polarity in exactly the same direction. The photons are emitted in all directions, so some will be traveling along the axis of the gas tube. Those will stimulate emission of photons traveling along the axis. Mirrors placed at either end of the tube allow those photons to be reflected back and forth, encountering excited neon atoms and stimulating more photons that will travel back and forth along the axis. Eventually, a large proportion of photons are moving along the axis de-exciting neon atoms while the electrons still being ejected from the cathode still are exciting helium atoms that re-excite neon atoms.

To use this highly coherent light source, it is sufficient to make one of the mirrors only 99 percent reflective, so that 1 percent is transmitted. That 1 percent is enough to read bar codes, help surgeons to position invisible infrared cutting beams, and help with precise length measurements. The photons used for these purposes make visible red light of 632.8-nm wavelength, though there are other neon transitions that can provide other wavelengths.

TECHNOLOGY AND CURRENT USES

Neon was the gas originally used for lighted signs, though many such signs that do not use neon still carry that name. Various gases are used

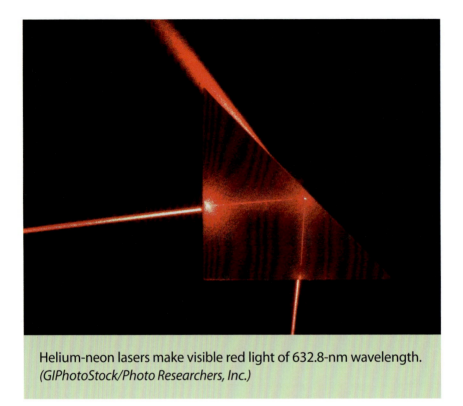

Helium-neon lasers make visible red light of 632.8-nm wavelength. *(GIPhotoStock/Photo Researchers, Inc.)*

in lighting fixtures. Neon, however, is often preferred for outdoor night-time lighting—especially in advertising—because the light emitted by neon is a brilliant red color that readily attracts attention. The human eye is especially sensitive to color in that part of the spectrum, making neon lights real attention-getters.

The gas that provides the important metastable state in helium-neon lasers is neon. The radiation that is stimulated and amplified is a result of de-excitation of neon atoms. Helium-neon lasers can be used to read bar codes, help surgeons to position invisible infrared cutting beams, and help with determining precise measurements.

A newer application is the use of neon gas for cooling purposes, as it is less expensive than liquid helium for the same applications. In electronics, it is used in lightning protection equipment, high voltage tubes, and wave meters.

7

Argon

Argon, element number 18, is formed by the radioactive decay of unstable potassium isotopes. Because potassium is an abundant element on Earth, sizable quantities of argon can be produced. In fact, argon comprises about 1 percent of Earth's atmosphere, making it the most abundant noble gas in the atmosphere and the third most abundant gas of all types in the atmosphere. Like helium and neon, argon has no chemistry. Therefore, its presence in the atmosphere was a mystery for many years. Unlike nitrogen and oxygen—which comprise four-fifths and one-fifth of the atmosphere, respectively, and can be removed from the atmosphere by chemical reactions—argon undergoes no such reactions. Chemists in the late 1700s successfully identified 99 percent of the gases in the atmosphere, but were baffled by the remaining 1 percent. Being unable to react this remaining component of air with

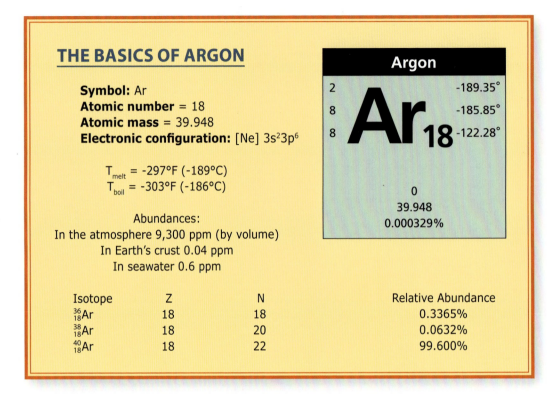

THE BASICS OF ARGON

Symbol: Ar
Atomic number = 18
Atomic mass = 39.948
Electronic configuration: [Ne] $3s^23p^6$

T_{melt} = -297°F (-189°C)
T_{boil} = -303°F (-186°C)

Abundances:
In the atmosphere 9,300 ppm (by volume)
In Earth's crust 0.04 ppm
In seawater 0.6 ppm

			Argon		
2					-189.35°
8		**Ar**			-185.85°
8			18		-122.28°
			0		
			39.948		
			0.000329%		

Isotope	Z	N	Relative Abundance
$^{36}_{18}$Ar	18	18	0.3365%
$^{38}_{18}$Ar	18	20	0.0632%
$^{40}_{18}$Ar	18	22	99.600%

any other chemicals prevented its identification for a century. Eventually, the problem was solved as chemists began to understand that a family of inert gases exists that resists undergoing chemical reactions. After techniques for liquefying air were developed in the 1890s, each of the inert gases found in air, including argon, were isolated and identified using spectroscopic methods. By determining the atomic weights of these gases, chemists were able to correctly place them in their own column following the halogens.

THE ASTROPHYSICS OF ARGON

Although argon makes up only a tiny fraction of our atmosphere, it is ubiquitous in the universe. Argon 40 is the most abundant isotope of argon in Earth's atmosphere, but argon 36 is the isotope most commonly observed in stellar spectra, including that of the Sun. That is because argon is an alpha-process element, meaning that it is generally synthesized by capture of a helium nucleus onto sulfur 32 as noted:

$$^{32}_{16}S + {}^{4}_{2}\alpha \rightarrow {}^{36}_{18}Ar.$$

This process occurs during the main sequence lifetime as a consequence of interactions in the star's core, as well as during supernovae explosions, which distribute argon into the interstellar medium.

Argon has been detected in the spectra of such diverse astronomical objects as our Sun, hot central stars of planetary nebulae, and white dwarfs. The most significant observations, however, may be those associated with new star formation in blue compact galaxies. The blue color of these galaxies indicates that most of the stars are very young and hot. Because these galaxies are relatively compact, the mass density is high, so collisions are many and frequent, and the high level of activity allows gas collections to gather matter to form new stars. These regions of the sky are important because they can provide the most reliable data on argon abundance in stars with low *metallicities*.

SCUBA DIVING AND ARGON

For cold-water scuba diving, dry suits (typically made of vulcanized rubber) may be worn to minimize loss of body heat. Worn over a synthetic undergarment, a dry suit itself is not thermally insulating, but

At left is the spectral tube, excited by a 5,000-volt transformer; at right is the argon spectrum as seen through a 600-line/mm diffraction grating. *(Department of physics and astronomy, Georgia State University)*

allows a layer of air or other insulating gas to be injected between it and the undergarment. The most viable gas for this purpose is argon.

To minimize heat transfer, a low thermal conductivity is required. At the molecular level, thermal conductivity depends directly on the *specific heat* of the gas and inversely on the mass and size of the atoms or molecules. This means that if the specific heat is low, the thermal conductivity will be low, and the larger the mass and diameter of a molecule the lower is the value of the conductivity.

The specific heat of a gas is high if it can hold a lot of heat energy. Molecules, with their vibrational and rotational states, are able to absorb energy in more ways than atoms, so monatomic gases have a lower specific heat than molecular gases. But atoms have smaller diameters than molecules, and so are likely to collide less frequently. Collisions within the gas allow the particles to slow down, and the gas remains cooler. Mass must also be taken into account, since heavier particles move more slowly than light ones.

The ideal gas for thermal insulation, therefore, would be either monatomic with large heavy atoms or molecular with a low specific heat. The noble gases of helium, neon, argon, krypton, xenon, and radon exist in nature as monatomic gases, so it would seem that the best insulator would be made of the heaviest noble gas, radon. Radon is radioactive, however, and so is not a good choice to have surrounding the human body. Krypton and xenon, the next heaviest, would work fine, but they are exorbitantly expensive, and so are not economically practical. That brings it down to the candidate of argon, whose thermal conductivity is 67 percent that of air.

Some molecular gases with rather low specific heats are also viable because of their constituents' heavier mass and size. Carbon dioxide, sulfur hexafluoride (SF_6), and Freon 14 (CF_4) are in the range of physical requirements, with thermal conductivities relative to air of 62 percent, 50 percent, and 62 percent, respectively. So they are better thermal insulators. The problems with using these gases are economic and physiological. The issue is chemical reactivity. The noble gases, like argon, generally do not react with other atoms and molecules, but the chemical species can and do present a multitude of problems to the human system.

FREEZING CANCER

A major problem surgeons face when removing cancerous tumors is trying to remove all of the malignant cells without harming the adjacent healthy cells. If all of the malignant cells are not completely removed, then the tumor is very likely to grow back again. Removing too many of the nearby healthy cells could lead to other health problems. One approach to destroying the malignant cells is to inject anticancer drugs. It is frequently the case, however, that the drugs do not reach every malignant cell, so that the treatment has to be repeated.

Recently, surgeons have begun freezing the malignant tissue using either liquid argon or liquid nitrogen. Argon gas is faster and therefore less harmful, as the tissue spends less time in a non-natural state. Using ultrasonic imaging, surgeons pinpoint where to inject the cryogenic liquid. Once the tissue is frozen, the anticancer drug can then be injected. Cancer cells absorb more of the drug when frozen than they would at normal temperatures. This means that lower dosages of the anticancer drugs can be administered, resulting in fewer side effects normally associated with chemotherapy. This combination of cryosurgery and chemotherapy can lead to complete destruction of a cancerous tumor without the necessity of a repeat procedure. To date, surgeons have been successful in treating tumors of the liver and prostate gland.

Surgeons can freeze malignant tissues using liquid argon. *(Colin Cuthbert/Photo Researchers, Inc.)*

Argon is heavy enough, nonreactive enough, and cheap enough to be the gas of choice for insulating divers against the chill of the sea.

TECHNOLOGY AND CURRENT USES

Because argon is chemically inert, there are no argon compounds. All of argon's uses are of the pure element. For example, argon is used in incandescent light bulbs and is the gas of choice for thermal insulation in dry suits used for cold-water scuba diving because it is heavy enough, nonreactive, and inexpensive. Arc welding requires shielding gases such as argon. Argon gas can be used in lasers and emits discrete frequencies of light in the visible and ultraviolet regions of the spectrum. Surgeons can use liquid argon to freeze cancerous tissue, which absorbs more anticancer drug when frozen than at normal temperatures, hastening treatment.

8

Krypton
and Xenon

Krypton and xenon are elements number 36 and 54, respectively. They are found in trace amounts in the atmosphere; air is liquefied and then distilled to collect each of its constituent gases individually. Extremely small amounts of krypton and xenon also occur in minerals and meteorites. Like all noble gases, on Earth they result from the radioactive decay of other elements.

Krypton and xenon are discussed together in this chapter because—unlike the noble gases helium, neon, and argon—the elements krypton and xenon do form chemical compounds. The difference in chemical behavior is a result of the larger sizes of krypton and xenon atoms. Krypton and xenon have valence electrons in "d" subshells; the "d" electrons are the ones capable of bonding with other elements, principally fluorine.

THE BASICS OF KRYPTON

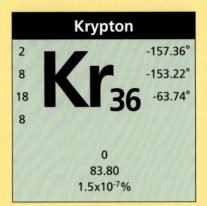

Symbol: Kr
Atomic number: 36
Atomic mass: 83.798
Electronic configuration: $[Ar]4s^23d^{10}4p^6$

T_{Melt} = -251°F (-157°C)
T_{Boil} = -244°F (-153°C)

Abundances:
In the atmosphere 1.14 ppm (by volume)
In Earth's crust: Negligible
In seawater: Negligible

Isotope	Z	N	Relative Abundance
$^{78}_{36}$Kr	36	42	0.35%
$^{80}_{36}$Kr	36	44	2.28
$^{82}_{36}$Kr	36	46	11.58%
$^{83}_{36}$Kr	36	47	11.49%
$^{84}_{36}$Kr	36	48	57.00%
$^{86}_{36}$Kr	36	50	17.30%

THE BASICS OF XENON

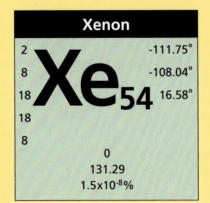

Symbol: Xe
Atomic number: 54
Atomic mass: 131.293
Electronic configuration: $[Kr]5s^24d^{10}5p^6$

T_{Melt} = -169°F (-112°C)
T_{Boil} = -162°F (-108°C)

Abundances:
In the atmosphere 0.086 ppm (by volume)
In Earth's crust: Negligible
In seawater: Negligible

Isotope	Z	N	Relative Abundance
$^{128}_{54}$Xe	54	74	1.92%
$^{129}_{54}$Xe	54	75	26.44%
$^{130}_{54}$Xe	54	76	4.08%
$^{131}_{54}$Xe	54	77	21.18%
$^{132}_{54}$Xe	54	78	26.89%
$^{134}_{54}$Xe	54	80	10.44%
$^{136}_{54}$Xe	54	82	8.87%

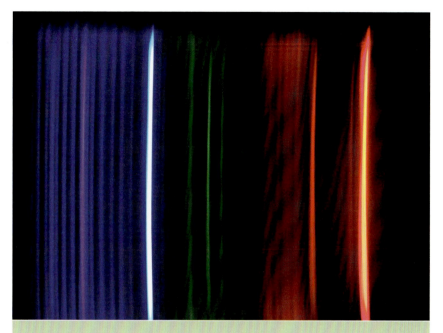

The krypton emission spectrum has its brightest line in the green region of visible light. *(Ted Kinsman/Photo Researchers, Inc.)*

THE ASTROPHYSICS OF KRYPTON AND XENON

Krypton and xenon, both being heavier than iron, are synthesized in supernovae via the rapid capture by iron nuclei of a succession of neutrons, which is called the r-process. Some fraction of the heavy noble gases also builds up slowly over thousands of years in the atmospheres of large-mass stars via neutron capture and electron release, with the requirement that iron 56 nuclei be available as seeds—remnants of prior supernova explosions. Because this synthesis proceeds relatively slowly due to a low density of neutrons, it is called the s-process.

While abundances of krypton and xenon are small, they are measurable, and some interesting anomalies have shown up in recent studies. Since these elements are chemically very weakly reactive and unlikely to form mechanical bonds, they should not have become incorporated into interstellar dust grains like most other elements. So their abundances should have remained constant over the billions of years since the formation of the solar system. Recent observations of spectra from the extrasolar but near-interstellar medium, however, show that significantly more krypton has collected in the solar system than in the vicinity around it.

Additionally, studies of other planetary atmospheres report varied abundances. The Martian atmosphere, for example, shows an interesting mix of xenon isotopes differing from that in Earth's atmosphere. Measurements of the atmosphere of Jupiter's moon, Titan, conclude a surprising depletion of xenon and krypton relative to Jupiter's, perhaps owing to trapping in cages formed by water molecules, called clathrate hydrates, which are now frozen on Titan's surface. Samples of isotopically strange xenon have been found in meteorites and lunar rocks. The field of noble gas astronomy continues to provide intriguing clues about the history of the solar system and evolution of the universe.

THE SPARSE CHEMISTRY OF KRYPTON—NOT SO FOR XENON

Neil Bartlett's (1932–2008) first synthesis of a noble gas compound was motivated by the reaction that occurs between the gas PtF_6 and oxygen gas to form a crystalline solid, $[O_2^+][PtF_6^-]$. Substituting xenon gas for oxygen, he obtained a complex red crystalline solid often represented as $[Xe^+][PtF_6^-]$. There are now known to be a number of compounds in which xenon is bonded to either fluorine or oxygen. In the solid state, these include XeF_2, XeF_4, XeF_6, and XeO_3. Gaseous compounds include XeO_4.

Krypton can be induced to form KrF_2, in analogy to XeF_2, by passing an electric discharge through a mixture of krypton and fluorine gases. The krypton difluoride formed is a volatile solid consisting of clear, colorless crystals. However, KrF_2 is relatively unstable compared with XeF_2. Repeated attempts have been made to form compounds of helium, neon, and argon, but any species produced tend to be only transient at low temperatures and completely unstable at room temperature. Because elements in the same column of the periodic table possess similar properties, chemists presume that radon exhibits at least as much chemistry as xenon does. The scarcity and radioactivity of radon, however, hampers its study.

DEFINING THE METER

During medieval times, units of length were based on the body dimensions of the reigning monarch. For example, one inch was the length of a digit of the king's finger. One foot was the length of the king's foot. One yard was the distance from the tip of the king's nose to the tips of his outstretched hand. In the 1790s, however, the French National Assembly adopted the metric system, which used the meter as the standard unit of length.

It was soon agreed upon that the meter needed a precise standard. The first standard that was proposed was that the meter was to be defined as one ten-millionth of the distance from the North Pole to the equator along the meridian that passed through the city of Paris. That distance was difficult to measure because of the flattening of Earth toward the equator, but it was used in the 1880s to cast a 1-m-long bar composed of a platinum-iridium alloy that was kept in a vault in Paris.

In 1960, that definition of a meter was replaced with a more precise standard based on a wavelength of radiation emitted by an electronic transition of krypton 86. In 1983, the krypton-based standard was replaced by the wavelength of a helium-neon laser. By the existing standard, 1 meter is currently defined as the distance traveled by light in a vacuum during a time interval of 1/299,792,458 of a second. For historical purposes, the platinum-iridium bar from the 1880s is still kept on display in Paris.

The universal standard meter iron bar is kept in Paris. *(ca. 1875–89, NIST Museum Collection)*

EXCIMER LASERS

While interactions by any of the noble gases with other elements are extremely rare in nature, they can be stimulated in the laboratory. For example, an excited state of krypton fluoride (KrF*) can be coaxed into existence by accelerating electrons through a mixed gas of krypton atoms and fluorine molecules (F_2). This diatomic molecule, called an excimer, exists only in the excited state; there is no ground state, so population inversion is easily achieved.

$$2 \; Kr + F_2 \rightarrow 2 \; KrF^*$$

Recombination with a free fluorine atom de-excites the excimer, producing laser light (γ) at the ultraviolet wavelength of 248 nm.

$$KrF^* + F \rightarrow Kr + F_2 + \gamma$$

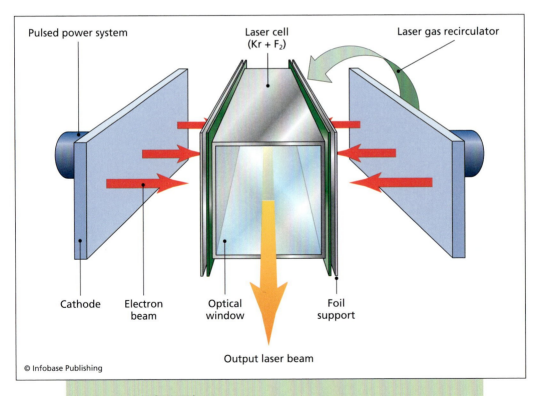

Pulsed power system   Laser cell (Kr + F_2)   Laser gas recirculator

Cathode   Electron beam   Optical window   Foil support

Output laser beam

© Infobase Publishing

Schematic of a KrF laser

The lasing action can then proceed as in the He-Ne laser, through the use of highly reflective mirrors at either end of the gas tube.

Excimer lasers produce power in brief pulses; this is accomplished by switching on and off the cathode power that provides the accelerated electron beam. Each pulse can provide energy of 1,000 joules or more. Since power is measured in joules per second, the shorter the pulses can be made, the more power is produced. Pulses of 20 nanoseconds are typical, but pulses as brief as 4 nanoseconds have been achieved, resulting in power ratings on the order of 200 terawatts. Such tremendous power is advancing fusion research. Direct-drive laser fusion, where the laser is used to heat fuel pellets, may prove more viable than efforts in inertial confinement fusion. The United States Naval Research Laboratory is a leader in this field.

Somewhat less powerful, yet useful, lasers use other noble gases with halogens to produce various wavelengths (γ) of laser light, as noted by the table below.

Excimer lasers have high beam uniformity compared with other lasers, and—perhaps most importantly—the beam spot size can be reduced and enlarged at will, so areas of different sizes can be targeted. This feature, combined with the short wavelengths and high power, makes them valuable in microchip manufacturing and laser eye surgery. The ultraviolet wavelengths, easily absorbed by human tissues, can be hazardous in terms of carcinogenic risks, but this ease of absorption has promise for medical research, especially in lipids and proteins.

LASER WAVELENGTH PRODUCED BY VARIOUS GASEOUS MEDIA

GAS	λ (NM)
argon fluoride	193
xenon fluoride	351
xenon chloride	308
krypton chloride	222

XENON AND EXOTIC PROPULSION

Exotic propulsion is the term used to describe methods that involve something other than the standard fuel-burning combustion that propels cars, airplanes, and rockets. The field is important to NASA and anyone interested in space travel beyond Mars.

The highest speed thus far attained by a human-made craft was reached by *Voyager 1.* Launched in 1977, the spacecraft was able to travel faster than 38,000 miles per hour because of a strategic trajectory that took advantage of Jupiter's considerable gravitational force as a one-time boost. The opportunity for such an augmentation is rarely available, however, and speeds many times greater are needed if travel to outer planets and other stars is to be accomplished within the time frame of one human generation.

Ion propulsion is a promising area of research that is being tested for this purpose. The basic physics is straightforward.

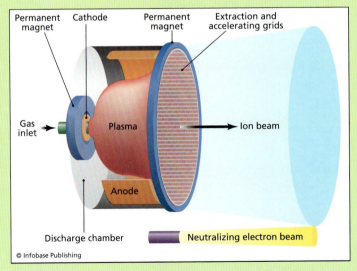

© Infobase Publishing

In ion propulsion, positively charged ions, as opposed to neutral atoms, are easily accelerated by an electric field, and can thus provide thrust.

A Boeing Delta II rocket launched the New Millennium Program *Deep Space 1* **spacecraft on October 24, 1998.** *(NASA)*

Force equals mass times acceleration. The force that propels a spacecraft (thrust) is equal to the mass of the spacecraft times its acceleration and is proportional to the speed of the exhaust gases. In rocket fuel combustion, a large amount of mass is burned quickly, which means a strong thrust and a rapid acceleration, but this requires the spacecraft to carry a great deal of fuel. This is a viable option for short hops, but not for long flights. For extended space exploration, a lightweight, constantly available source of acceleration will be needed.

Positively charged ions, as opposed to neutral atoms, are easily accelerated by an electric field, and can thus provide thrust. This requires only a battery. The ions are created by sending electrons into a chamber to collide with a neutral gas. Xenon is a good choice; nonreactive by nature, it remains pure in stor-

(continues)

(continued)

age for very long periods of time. (Krypton would also serve, but weighs more.) Since the ions are low mass, the thrust they provide is small, so the craft cannot pick up speed quickly. The process is continuous, however, and the constant increase in velocity over a period of years provides impressive results.

In October 1998 NASA used ion propulsion engines on *Deep Space 1*—a mission to intersect a comet passing 200 million miles (321.9 million km) from the Sun. On September 27, 2007, a NASA probe called *Dawn* launched and is currently employing ion power on a mission to the asteroid belt, which lies between the orbits of Mars and Jupiter. With its propulsion system, it could reach a speed of 24,000 miles per hour (38,624.3 km/hr) in five years. Solar panels keep the battery charged to create the potential difference needed to accelerate the electrons and xenon ions in the system.

TECHNOLOGY AND CURRENT USES

Although compounds of krypton and xenon have been made, few, if any, uses are known for them. The uses of krypton and xenon are of the pure elements.

Krypton-argon mixtures are used in fluorescent and tungsten light bulbs. Electric-arc lamps filled with krypton are used to mark airport runways at night. Krypton is the most common noble gas used in excimer lasers.

Radioactive isotopes of krypton can be used in lamps for illumination based on the radiation from the radioactive decay activating a phosphor coating. Xenon-filled arc lamps are particularly intense and are used to project motion pictures. Xenon is used in photographic flash bulbs; one bulb can be used up to 10,000 times before burning out.

Energy-efficient windows may use krypton or a krypton-argon mixture between double panes for a high insulation rating.

A mixture of 20 percent oxygen and 80 percent xenon has deep anesthetic qualities and can be used during surgery. Xenon is completely nontoxic and nonflammable, so it poses none of the dangers of more conventional anesthetics such as ether or ethylene.

Liquid xenon can be used in bubble chambers to study the production and properties of subatomic particles. Xenon gas is used in neutron counters, X-ray detectors, ionization chambers for cosmic rays, and arc lamps that produce ultraviolet radiation.

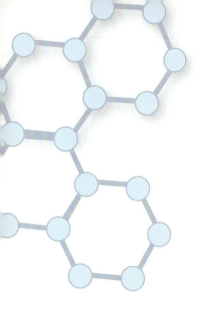

9

Radon:
A Common Menace

Radon is element number 86. As all of radon's isotopes have short half-lives, they tend to decay rather quickly. Because there is so little radon available for study at any one time, relatively little is known about its chemistry. Of greater interest are radon's radioactive decay properties. Radon's three longest-lived isotopes are Rn-219 ($t_{1/2}$ = 3.96 seconds), Rn-220 ($t_{1/2}$ = 55.6 seconds), and Rn-222 ($t_{1/2}$ = 3.8 days). Each of these isotopes is an emitter of alpha particles, as illustrated, for example, by the decay of Rn-222:

$$^{222}_{86}Rn \rightarrow {}^{4}_{2}He + {}^{218}_{84}Po,$$

where $^{4}_{2}He$ is an alpha particle.

THE GEOLOGY OF RADON

Most of the radon on Earth comes not from the stars, but from the spontaneous radioactive decay of terrestrial uranium 238 and its daugh-

THE BASICS OF RADON

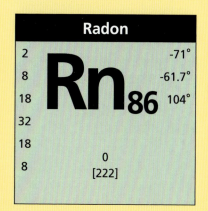

Symbol: Rn
Atomic number: 86
Atomic mass: 222.0176
   (of longest-lived isotope ^{222}Rn)
Electronic configuration:
   [Xe]$6s^2 4f^{14} 5d^{10} 6p6$

T_{Melt} = -96°F (-71°C)
T_{Boil} = -79°F (-62°C)

Abundance: Negligible

Note: There are more than 30 known radioactive isotopes of radon, all short-lived.

ter products, protactinium (Pa), thorium (Th), and radium (Ra), in the following sequence.

$$^{238}U \rightarrow {}^{234}Th + \alpha$$

$$^{234}Th \rightarrow {}^{234}Pa + \beta$$

$$^{234}Pa \rightarrow {}^{234}U + \beta$$

$$^{234}U \rightarrow {}^{230}Th + \alpha$$

$$^{230}Th \rightarrow {}^{226}Ra + \alpha$$

$$^{226}Ra \rightarrow {}^{222}Rn + \alpha$$

Although an average value for the concentration of uranium in ground rock and soil is about 2 ppm, it can be much higher where geological conditions are optimal, especially in granite, volcanic rock, and ancient

Geiger counters can be used to detect alpha radiation from radon decay. (AP/The Saginaw News, Jeff Schrier)

shale—as in western Ohio, where the uranium content commonly reaches twice the average value.

Uranium and radium are solids and are locked up in the rock they inhabit, but radon's natural state is gaseous. Upon the fission of a stationary ^{226}Ra nucleus, an alpha particle and radon nucleus are emitted in opposite directions, so there is a chance for the radon to escape from the solid part of the rock into a porous region, which happens with about a 30-percent probability. It is then easy for the radon to travel along the pores until it escapes into the atmosphere.

Radon is radioactive with a fairly short half-life (only 3.8 days), so the portion that does not escape from the rock soon decays by a chain of alpha and beta emissions to lead 206, which is stable. All of the radioactive transitions described here have little or no effect on humans unless the escaping radon gas is channeled into human habitations through their foundations.

RADON AND HYDROLOGY

Of all the noble gases, radon is the most soluble in water. If the pores in ground rock are full of water, radon gas travels more slowly than in air

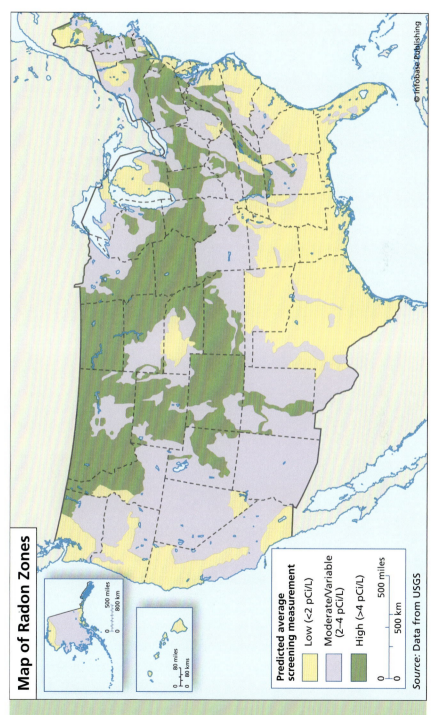

Map of Radon Zones

Predicted average screening measurement

Low (<2 pCi/L)

Moderate/Variable (2–4 pCi/L)

High (>4 pCi/L)

Source: Data from USGS

© Infobase Publishing

Radon map of U.S. areas in green have the highest risk of exposure while yellow areas have the lowest.

BAD AIR: IN BASEMENTS AND STONE HOUSES

Radon inhaled in small concentrations is not dangerous to humans, but in certain circumstances it can exceed a safe level. Radon gas seeping up from underground can preferentially enter the foundation of a house on its way to the surface, rather than continuing around the house to exit into the surrounding air. Any gas moves more freely in areas of lower pressure, where it will suffer fewer collisions with other molecules. Atmospheric pressure above the ground is lower than the pressure in the soil,

Houses made of rock or stone may have a higher radon content. *(minik/Shutterstock)*

and may end up in rivers, lakes, and *aquifers*. The highest concentrations accumulate in aquifers, which are embedded in rock or soil, so it is more difficult for the radon gas to escape. Radon can, therefore, be found in public water supplies and in water from private wells. Household water delivered by public utilities undergoes a series of treatments where it is exposed to air, so very little reaches the tap. Private wells are another issue, and may contain considerable dissolved radon that

and radon gas will reach a house's foundation before it encounters the soil surface. Generally, the earth around a foundation has been disturbed and is therefore more porous, providing more freedom of motion for the atoms in the gas. If the foundation were hermetically sealed, the radon could not get in. Virtually all home foundations, however, have cracks (even if only hairline) due to settling of the earth. Additionally, crawl spaces are often intentionally built for purposes such as examining the plumbing. So radon gas is often found in higher concentrations in homes than in the atmosphere, especially in basements.

The health risks come from the alpha particles emitted during the spontaneous decay of the radon atoms. In this case, a short half-life is a problem because the radon atom has little chance of escape from the home through windows, doors, or cracks before it fissions. Alpha particles carry charge. Any traveling charged particle can be classified as ionizing radiation, which means that it can remove an electron from an atom, molecule, or cell. This is a carcinogenic process in the human body, damaging DNA even when the radiation does not invade the nucleus of the cell. The lungs are particularly susceptible, as the gas is inhaled during normal breathing.

Radon is colorless and odorless, but can be detected by ion-monitoring devices that are easily available in hardware stores. Some state health departments offer free test kits to residents.

escapes into the air of the home, especially during showers or the heating of water for cooking. These households are at risk of an unhealthy fraction of radon in the air and should have the water tested.

TECHNOLOGY AND CURRENT USES

Radon's several uses result from its radioactive nature. Radon has been used in the treatment of cancer, although today other radioactive sources are more likely to be used. When mixed with beryllium, radium is used as a source of neutrons. Monitoring of changing radon concentrations in springs, well water, soil, and rocks is being investigated as a means of predicting earthquakes. Radon is also used as a tracer gas in studying large-scale motion of air masses across continents and the oceans.

10

Conclusions and Future Directions

The periodic table of the elements is a marvelous tool that scientists have only begun to investigate. The key to its utility is its organization, the patterns it weaves, and how it can guide the eye and the mind to understand how far science has come and where human knowledge can lead.

SPECULATIONS ON FUTURE DEVELOPMENTS

It is important for scientists to think ahead, to attempt to guess what areas are ripe for investigation and which may be bound for oblivion. If one considers recent remarkable leaps in information science, medicine, and particle physics that have materialized in the past century, it is clear that predictions are bound to be less impressive than eventualities, but there are some obvious starting points. Some of those, especially related to the halogens and noble gases, are suggested here.

NEW PHYSICS

As described throughout this volume, the astrophysics of most of the halogens and noble gases requires further investigation. Many more stellar spectra need to be studied in order to understand how important the contribution of asymptotic giant branch (AGB) star activity is to fluorine abundance in interstellar space. Bromine is another mystery; observational evidence is scarce, and its stellar evolution is confusing. The calculation of neon abundance in stars will also require further effort, especially in order to understand solar convection: Calculations of the behavior of gases in the Sun's convection zones do not agree with the accepted value of neon abundance. Other important areas of exploration in astrophysical neon research include its content in meteorites and in preplanetary solar system disks. Further exploration regarding chlorine, krypton, and xenon abundance in interstellar space as well as planetary atmospheres will help clarify how stars and planets evolve.

On Earth, there is plenty of scientific work to look forward to regarding research and societal needs. The imminent helium shortage will require attention, and fluorine supplementation in water supplies warrants additional investigation. In the medical arena, further understanding of bromine's radioactive decays, purification methods for [211]At samples, and the effect of changing human iodine intake will be important areas of research.

NEW CHEMISTRY

While the discovery of element 117 has not yet been reported, it is reasonable to assume that it will be discovered fairly soon. Since it will lie below astatine in the periodic table, it should therefore be a halogen and a solid, like iodine and astatine. If chemists succeed in synthesizing weighable quantities of element 117, it should also prove to be the most metallic of the halogens, since the trend is for the metallic nature of elements to increase upon descending a column.

In 2006, scientists succeeded in synthesizing a few atoms of element 118; it lies below radon in the periodic table, which places it in the family of noble gases. A general trend in the noble gas family, however, is for the boiling points of the elements to increase as the molecular weights increase. As a consequence of this trend, element 118 should be a liquid at room temperature and one atmosphere pressure, not a gas, thus potentially making element 118 the first "noble liquid." Unfortunately, the extremely short half-life (less than one-thousandth of a second) of the known isotope of element 118 currently prevents the synthesis of quantities large enough for the element's chemical and physical properties to be studied. It could be years before research reveals more.

The properties of the halogens have been studied so extensively for so many years that it is unlikely that many more uses will be found for the elements themselves or for inorganic compounds of these elements. Fluorine, chlorine, bromine, and iodine are often added to organic compounds, however, so that possibly useful new substances are waiting to be discovered.

Ever since Neil Bartlett's initial discovery in 1962 of the first chemical compound that contained xenon, extensive experiments have been conducted in the search for novel noble gas compounds. Since almost a half century has elapsed since Bartlett's pioneering work—and uses have not yet been found for the compounds made so far—it is beginning to look unlikely that any uses of noble gas compounds will be discovered.

The unlikelihood of discovering any novel halogen or noble gas compounds in the near future should not be interpreted to denigrate the importance of these two families of elements. Their prominent role in the world already ensures their continued importance. Disinfectants, sterilants, fluorides in drinking water and toothpaste, commercially important plastics, bleach and other household cleansers, insecticides, flame retardants, photographic film, radioisotopes in medical diagnosis and treatment, synthetic dyes, helium-filled balloons, cryogenics, neon lights, lasers, and light bulbs will continue to play prominent roles in the economy and in everyday life.

Many of these everyday substances, however, present associated hazards to human health and environmental welfare. Chlorine compounds are particularly challenging. *Polychlorinated biphenyls,* for example—which were used for decades in paints, sealants, pesticides, and electronics and are now outlawed because of their toxicity—still persist in the environment. Research must proceed regarding novel ways to remove these toxins from the water system. Polyvinyl chloride (PVC) also has many uses, including applications in plumbing, vinyl siding, magnetic stripe cards, upholstery, and flooring. It has been shown, however, that more than 100 volatile organic compounds may be released into the atmosphere from many of these products. Ingestion or inhalation of these chemicals can lead to problems, particularly in reproductive health. Likewise, chlorofluorocarbons, which break down the ozone layer, must be eradicated, and human production of greenhouse gases like SF_6 will need to be curtailed. The future of the science of halogens may well need to be deeply invested in finding solutions to problems that scientists created.

SI Units and Conversions

UNIT	QUANTITY	SYMBOL	CONVERSION
Base units			
meter	length	m	1 m = 3.2808 feet
kilogram	mass	kg	1 kg = 2.205 pounds
second	time	s	
ampere	electric current	A	
kelvin	thermodynamic temperature	K	1 K = 1°C = 1.8°F
candela	luminous intensity		
mole	amount of substance	mol	
Supplementary units			
radian	plane angle	rad	pi / 2 rad = 90°
steradian	solid angle	sr	
Derived units			
coulomb	quantity of electricity	C	
cubic meter	volume	m^3	1 m^3 = 1.308 $yards^3$
farad	capacitance	F	
henry	inductance	H	
hertz	frequency	Hz	
joule	energy	J	1 J = 0.2389 calories
kilogram per cubic meter	density	$kg\ m^{-3}$	1 $kg\ m^{-3}$ = 0.0624 $lb.\ ft^{-3}$
lumen	luminous flux	lm	
lux	illuminance	lx	
meter per second	speed	$m\ s^{-1}$	1 $m\ s^{-1}$ = 3.281 $ft\ s^{-1}$

UNIT	QUANTITY	SYMBOL	CONVERSION
meter per second squared	acceleration	m s^{-2}	
mole per cubic meter	concentration	mol m^{-3}	
newton	force	N	1 N = 7.218 lb. force
ohm	electric resistance	Ω	
pascal	pressure	Pa	1 Pa = $\dfrac{0.145 \text{ lb}}{\text{in}^{-2}}$
radian per second	angular velocity	rad s^{-1}	
radian per second squared	angular acceleration	rad s^{-2}	
square meter	area	m^2	1 m^2 = 1.196 yards^2
tesla	magnetic flux density	T	
volt	electromotive force	V	
watt	power	W	1W = 3.412 Btu h^{-1}
weber	magnetic flux	Wb	

PREFIXES USED WITH SI UNITS		
PREFIX	**SYMBOL**	**VALUE**
atto	a	× 10^{-18}
femto	f	× 10^{-15}
pico	p	× 10^{-12}
nano	n	× 10^{-9}
micro	μ	× 10^{-6}
milli	m	× 10^{-3}
centi	c	× 10^{-2}
deci	d	× 10^{-1}
deca	da	× 10
hecto	h	× 10^{2}
kilo	k	× 10^{3}
mega	M	× 10^{6}
giga	G	× 10^{9}
tera	T	× 10^{12}

List of Acronyms

AGB	Asymptotic giant branch
CFC	Chlorofluorocarbon
DDT	Dichlorodiphenyltrichloroethane
DNA	Deoxyribonucleic acid
GHG	Greenhouse gas
HCFC	Hydrochlorofluorocarbon
HR	Hertzsprung-Russell
ISM	Interstellar medium
LANL	Los Alamos National Laboratory
NASA	National Aeronautics and Space Administration
PET	Positron emission tomography
PP	Proton-proton
PVC	Polyvinyl chloride
T3	Tri-iodothyronine
T4	Thyroxine
UV	Ultraviolet
VOC	Volatile organic compounds
WR	Wolf-Rayet

Periodic Table of the Elements

Periodic Table of the Elements

Legend (example cell):
- 3 — Atomic number
- Li — Symbol
- 6.941 — Atomic weight

Color key:
- Halogens
- Metals
- Nonmetals
- Metalloids
- Unknown

1 IA	2 IIA	3 IIIB	4 IVB	5 VB	6 VIB	7 VIIB	8 VIIIB	9 VIIIB	10 VIIIB	11 IB	12 IIB	13 IIIA	14 IVA	15 VA	16 VIA	17 VIIA	18 VIIIA
1 H 1.00794																	2 He 4.0026
3 Li 6.941	4 Be 9.0122											5 B 10.81	6 C 12.011	7 N 14.0067	8 O 15.9994	9 F 18.9984	10 Ne 20.1798
11 Na 22.9898	12 Mg 24.3051											13 Al 26.9815	14 Si 28.0855	15 P 30.9738	16 S 32.067	17 Cl 35.4528	18 Ar 39.948
19 K 39.0938	20 Ca 40.078	21 Sc 44.9559	22 Ti 47.867	23 V 50.9415	24 Cr 51.9962	25 Mn 54.938	26 Fe 55.845	27 Co 58.9332	28 Ni 58.6934	29 Cu 63.546	30 Zn 65.409	31 Ga 69.723	32 Ge 72.61	33 As 74.9216	34 Se 78.96	35 Br 79.904	36 Kr 83.798
37 Rb 85.4678	38 Sr 87.62	39 Y 88.906	40 Zr 91.224	41 Nb 92.9064	42 Mo 95.94	43 Tc (98)	44 Ru 101.07	45 Rh 102.9055	46 Pd 106.42	47 Ag 107.8682	48 Cd 112.412	49 In 114.818	50 Sn 118.711	51 Sb 121.760	52 Te 127.60	53 I 126.9045	54 Xe 131.29
55 Cs 132.9054	56 Ba 137.328	57–70 ☆	72 Hf 178.49	73 Ta 180.948	74 W 183.84	75 Re 186.207	76 Os 190.23	77 Ir 192.217	78 Pt 195.08	79 Au 196.9655	80 Hg 200.59	81 Tl 204.3833	82 Pb 207.2	83 Bi 208.9804	84 Po (209)	85 At (210)	86 Rn (222)
87 Fr (223)	88 Ra (226)	89–102 ★	104 Rf (261)	105 Db (262)	106 Sg (266)	107 Bh (262)	108 Hs (263)	109 Mt (268)	110 Ds (271)	111 Rg (272)	112 Cn (277)	113 Uut (284)	114 Uuq (285)	115 Uup (288)	116 Uuh (292)	117 Uus	118 Uuo (294)

☆ Lanthanides

57 La 138.9055	58 Ce 140.115	59 Pr 140.908	60 Nd 144.24	61 Pm (145)	62 Sm 150.36	63 Eu 151.966	64 Gd 157.25	65 Tb 158.9253	66 Dy 162.500	67 Ho 164.9303	68 Er 167.26	69 Tm 168.9342	70 Yb 173.04	71 Lu 174.967

★ Actinides

89 Ac (227)	90 Th 232.0381	91 Pa 231.036	92 U 238.0289	93 Np (237)	94 Pu (244)	95 Am 243	96 Cm (247)	97 Bk (247)	98 Cf (251)	99 Es (252)	100 Fm (257)	101 Md (258)	102 No (259)	103 Lr (260)

Numbers in parentheses are atomic mass numbers of most stable isotopes.

© Infobase Publishing

116

Element Categories

Element Categories

Nonmetals
1	H	Hydrogen
6	C	Carbon
7	N	Nitrogen
8	O	Oxygen
15	P	Phosphorus
16	S	Sulfur
34	Se	Selenium

Halogens
9	F	Fluorine
17	Cl	Chlorine
35	Br	Bromine
53	I	Iodine
85	At	Astatine

Noble Gases
2	He	Helium
10	Ne	Neon
18	Ar	Argon
36	Kr	Krypton
54	Xe	Xenon
86	Ra	Radon

Metalloids
5	B	Boron
14	Si	Silicon
32	Ge	Germanium
33	As	Arsenic
51	Sb	Antimony
52	Te	Tellurium
84	Po	Polonium

Alkali Metals
3	Li	Lithium
11	Na	Sodium
19	K	Potassium
37	Rb	Rubidium
55	Cs	Cesium
87	Fr	Francium

Alkaline Earth Metals
4	Be	Beryllium
12	Mg	Magnesium
20	Ca	Calcium
38	Sr	Strontium
56	Ba	Barium
88	Ra	Radium

Post-Transition Metals
13	Al	Aluminum
31	Ga	Gallium
49	In	Indium
50	Sn	Tin
81	Tl	Thallium
82	Pb	Lead
83	Bi	Bismuth

Transactinides
104	Rf	Rutherfordium
105	Db	Dubnium
106	Sg	Seaborgium
107	Bh	Bohrium
108	Hs	Hassium
109	Mt	Meitnerium
110	Ds	Darmstadtium
111	Rg	Roentgenium
112	Cn	Copernicium
113	Uut	Ununtrium
114	Uuq	Ununquadium
115	Uup	Ununpentium
116	Uuh	Ununhexium
118	Uuo	Ununoctium

Transition Metals
21	Sc	Scandium	39	Y	Yttrium	72	Hf	Hafnium
22	Ti	Titanium	40	Zr	Zirconium	73	Ta	Tantalum
23	V	Vanadium	41	Nb	Niobium	74	W	Tungsten
24	Cr	Chromium	42	Mo	Molybdenum	75	Re	Rhenium
25	Mn	Manganese	43	Tc	Technetium	76	Os	Osmium
26	Fe	Iron	44	Ru	Ruthenium	77	Ir	Iridium
27	Co	Cobalt	45	Rh	Rhodium	78	Pt	Platinum
28	Ni	Nickel	46	Pd	Palladium	79	Au	Gold
29	Cu	Copper	47	Ag	Silver	80	Hg	Mercury
30	Zn	Zinc	48	Cd	Cadmium			

Note: The organization of periodic table of the elements, while useful to chemists and physicists, may be confusing to nonscientists in that some groupings of similar elements appear as vertical columns (halogens, for example), some as horizontal rows (lanthanides, for example), and some as a combination of both (nonmetals).

The table of element categories is intended as a quick reference sheet to easily determine which elements belong to which groups. (Element 117 does not appear in this list because it is undiscovered as of the publishing of this book.)

Lanthanides
57	La	Lanthanum	62	Sm	Samarium	67	Ho	Holmium
58	Ce	Cerium	63	Eu	Europium	68	Er	Erbium
59	Pr	Praseodymium	64	Gd	Gadolinium	69	Tm	Thulium
60	Nd	Neodymium	65	Tb	Terbium	70	Yb	Ytterbium
61	Pm	Promethium	66	Dy	Dysprosium	71	Lu	Lutetium

Actinides
89	Ac	Actinium	94	Pu	Plutonium	99	Es	Einsteinium
90	Th	Thorium	95	Am	Americium	100	Fm	Fermium
91	Pa	Protactinium	96	Cm	Curium	101	Md	Mendelevium
92	U	Uranium	97	Bk	Berkelium	102	No	Nobelium
93	Np	Neptunium	98	Cf	Californium	103	Lr	Lawrencium

Chronology

1731 English chemist Henry Cavendish is born on October 10 in Nice, France.

1742 Swedish chemist Carl Wilhelm Scheele is born on December 9 in Stralsund, Sweden.

1743 French chemist Antoine Lavoisier is born on August 26 in Paris.

1745 Swedish chemist Johan Gottlieb Gahn is born on August 19 in Voxna, near Söderhamn.

1748 French chemist Claude Louis Berthollet is born on December 9 in Paris.

1774 Carl Scheele isolates chlorine gas.

Johan Gahn isolates manganese.

1777 French chemist Bernard Courtois is born on February 8 in Dijon, France.

1778 French chemist Joseph Louis Gay-Lussac is born on December 6 in Haute Vienne, France.

English chemist Humphrey Davy is born on December 17 in Penzance, England.

1785 Claude Louis Berthollet discovers aqueous chlorine's bleaching properties.

Henry Cavendish recognizes that air contains more than nitrogen and oxygen, but could not identify the remaining 1 percent (which is now known to be argon).

1786 Carl Wilhelm Scheele dies on May 26 in Köping, Sweden.

1789 Antoine Lavoisier's *Elementary Treatise of Chemistry* introduces a new definition of an element as a substance that cannot be broken down into simpler substances.

1794   Antoine Lavoisier is executed by French revolutionists on May 8 in Paris.

1795   France officially adopts the metric system.

1802   French chemist Antoine-Jerome Balard is born on September 30 in Montpellier, France.

1803   German chemist Carl Lowig is born on March 17 in Bad Kreuznach, Germany.

1810   Humphrey Davy claims chlorine is an element, not a compound, and names it.

    Henry Cavendish dies on February 24 in London.

1811   Bernard Courtois discovers the presence of iodine in kelp.

1813   Nicolas Clement, Humphrey Davy, and Joseph Louis Gay-Lussac announce the discovery of iodine. Bernard Courtois is given credit for priority of discovery.

1818   Johan Gottlieb Gahn dies on December 8 in Stockholm, Sweden.

1822   Claude Louis Berthollet dies on November 6 in Paris.

1824   French astronomer Pierre-Jules-César Janssen is born on February 22 in Paris.

1826   Antoine-Jerome Balard isolates bromine.

1829   Humphrey Davy dies on May 29 in Geneva, Switzerland.

1834   Russian chemist Dmitri Mendeleev is born on February 8 in Tobolsk, Siberia.

1836   English astronomer Sir Joseph Norman Lockyer is born on May 17 in Rugby, England.

1838   Bernard Courtois dies on September 27 in Paris.

1842   English physicist John William Strutt, the third Lord Rayleigh, is born on November 12 in Langford Grove, Essex, England.

1848   German physicist Friedrich Ernst Dorn is born on July 27 in Guttstadt, Prussia.

1850 Belgian chemist Paulin Louyet dies attempting to isolate fluorine.

Joseph Louis Gay-Lussac dies on May 9 in Paris.

1852 French chemist Ferdinand-Frederic-Henri Moissan is born on September 28 in Paris.

Scottish chemist Sir William Ramsay is born on March 15 in Glasgow, Scotland.

1868 Pierre Janssen and Joseph Lockyer discover spectral lines due to helium in the Sun's spectrum.

1869 Dmitri Mendeleev publishes the periodic table.

1870 French chemist Georges Claude is born on September 24 in Paris.

1872 English chemist Morris William Travers is born on January 24 in London.

1876 Antoine-Jerome Balard dies on March 30 in Paris.

1886 Henri Moissan successfully isolates fluorine.

1890 Carl Lowig dies on March 27 in Breslau, Germany.

1894 William Ramsay and Lord Rayleigh announce the discovery of argon.

1895 William Ramsay discovers helium in uranium ores.

1898 William Ramsay, Lord Rayleigh, and Morris Travers discover neon, krypton, and xenon.

1900 Friedrich Dorn discovers radon.

1904 William Ramsay is awarded the Nobel Prize in chemistry; Lord Rayleigh receives the Nobel Prize in physics.

1905 Italian-American physicist Emilio Segrè is born on February 1 in Rome, Italy.

1906 Henri Moissan receives the Nobel Prize in chemistry.

1907 Dmitri Mendeleev dies on February 2 in St. Petersburg, Russia.

Henri Moissan dies on February 20 in Paris.

Pierre Janssen dies on December 23 in Paris.

1910 William Ramsay establishes that radon is the heaviest known noble gas.

Georges Claude unveils the first neon lamp on December 11.

1916 William Ramsay dies on July 23 in Haslemere, Surrey, England.

Friedrich Dorn dies on December 16 in Saxony, Germany.

1919 Lord Rayleigh dies on June 30 in Terling Place, Essex, England.

1920 Joseph Lockyer dies on August 16 in Devon, England.

1924 The USS *Shenandoah* makes the first transcontinental flight of an airship filled with helium.

1932 British-American chemist Neil Bartlett is born on September 15 in Newcastle-upon-Tyne, England.

1938 Dr. Roy Plunkett invents Teflon® at the DuPont research laboratories.

1940 Astatine is discovered by a team of University of California, Berkeley, scientists led by Emilio Segrè.

1945 Grand Rapids, Michigan, becomes the first community in the world to modify its fluoride levels in drinking water to benefit dental health.

1960 Georges Claude dies on May 23 in Saint-Cloud, France.

1961 Morris Travers dies on August 25 in Stroud, Gloucestershire, England.

1962 Neil Bartlett prepares first compound of a noble gas, XeF_2.

1989 Emilio Segrè dies on April 22 in Lafayette, California.

2006 Researchers at the Lawrence Livermore National Laboratory in California and the Joint Institute for Nuclear Research in Dubna, Russia, report discovery of element 118, which is located just below radon in the periodic table.

2007   Former U.S. vice president Al Gore shares Nobel Peace Prize for raising awareness about the hazards of greenhouse gases.

2008   Neil Bartlett dies on August 5 in Walnut Creek, California.

2009   Physics Nobel Prize winner, Steven Chu, becomes U.S. secretary of energy.

Glossary

acid   a type of compound that contains hydrogen and dissociates in water to produce hydrogen ions.

actinide   the elements ranging from thorium (atomic number 90) to lawrencium (number 103); they all have two outer electrons in the 7s subshell plus increasingly more electrons in the 5f subshell.

alkali metal   the elements in column IA of the periodic table (exclusive of hydrogen); they all are characterized by a single valence electron in an s subshell.

alkaline earth metal   the elements in column IIA of the periodic table; they all are characterized by two valence electrons that fill an s subshell.

alkyl   in a molecule, a subgroup of carbon and hydrogen atoms with the general formula $-C_nH_{2n+1}$.

alkyl group   a group of carbon and hydrogen atoms with a general formula of C_nH_{2n+1} that is attached to another organic compound.

allotropic form (or allotropes)   two or more different forms of an element; the difference could be due to different numbers or arrangements of the atoms in molecules or due to different crystal structures.

alpha decay   a mode of radioactive decay in which an alpha particle is emitted. The daughter isotope has an atomic number two units less than the atomic number of the parent isotope, and a mass number that is four units less.

alpha particle   a nucleus of helium 4.

anaphylactic shock   a severe, potentially fatal, allergic reaction to foods, medicines, bee stings, and other substances.

anion   an atom with one or more extra electrons, giving it a net negative charge.

anode   in an electrochemical cell, the electrode at which an oxidation reaction occurs; in a battery, the negative electrode.

123

antiepileptic   a drug designed to treat the symptoms of epilepsy, for example, seizures.

antiparticle   a subatomic particle that has the same mass but opposite charge as the partner particle.

aqueous   describing a solution in water.

aquifer   an underground body of water surrounded by rock or soil.

asymptotic giant branch (AGB)   an area of the Hertzsprung-Russell diagram above the main sequence line, where some high mass stars (AGB stars) are mapped for luminosity and temperature.

atom   the smallest part of an element that retains the element's chemical properties; atoms consist of protons, neutrons, and electrons.

atomic mass   the mass of a given isotope of an element—the combined masses of all its protons, neutrons, and electrons.

atomic number   the number of protons in an atom of an element.

atomic weight   the mean weight of the atomic masses of all the atoms of an element found in a given sample, weighted by isotopic abundance.

autooxidation   (See **disproportionation.**)

bactericidal   a chemical or biological agent that can prevent infection by inhibiting the growth or action of microorganisms.

base   a substance that reacts with an acid to give water and a salt; a substance that, when dissolved in water, produces hydroxide ions.

beta decay   a mode of radioactive decay in which a beta particle—an ordinary electron—is emitted; the daughter isotope has an atomic number one unit greater than the atomic number of the parent isotope, but the same mass number.

binary compound   a chemical compound consisting of only two elements, but variable numbers of atoms of each of those elements.

brachytherapy   injection of a small source or "seed" to a specific physiological site.

brominating agent   a halogenating agent in which bromine atoms are transferred from one compound or ion to another compound or ion.

buoyancy the upward force exerted on an object that is either floating or submersed in a fluid.

carcinogen a chemical substance that causes cancer.

cathode in an electrochemical cell, the electrode at which a reduction reaction occurs; in a battery, the positive electrode.

cation an atom that has lost one or more electrons to acquire a net positive charge.

chemical change a change in which one or more chemical elements or compounds form new compounds; in a chemical change, the names of the compounds change.

chemotherapy a treatment for cancer in which patients are administered certain types of drugs.

chlorofluorocarbon a compound that contains carbon, chlorine, and fluorine; historically used in refrigeration systems and as a propellant in spray cans.

clathrate hydrate a molecule in which water forms a cage around an element.

cloud seeding the attempt to modify the weather by dispersing salts into the clouds so as to increase precipitation.

complex ion any ion that contains more than one atom.

compound a pure chemical substance consisting of two or more elements in fixed, or definite, proportions.

convection the mixing of fluid or gas by the effect of warm material rising and cool material falling.

cosmic ray particles that enter Earth's atmosphere from outer space.

covalent bond a chemical bond formed by sharing valence electrons between two atoms (in contrast to an ionic bond, in which one or more valence electrons are transferred from one atom to another atom).

critical exemption permission to use an otherwise banned chemical in situations in which its uses are essential, or critical, to human health; an example is the continued use of CFCs in inhalers for asthma sufferers, even though CFCs have been banned for other applications.

critical point temperature   the temperature of a pure substance at which the liquid and gaseous phases become indistinguishable.

cryogenic   the production of very low temperatures and the study of the properties of materials at low temperatures.

cyclotron   a circular accelerator of light charged particles.

dephlogisticated air   air that does not contain phlogiston.

diatomic molecule   a molecule that contains two atoms.

disproportionation   an oxidation-reduction reaction in which some atoms of an element are oxidized while the remaining atoms of the same element are reduced.

Doppler shift   the phenomenon of apparent increase or decrease in frequency of sound or light as a result of the velocity of the source.

ductility   the ability of certain metals to be able to be drawn into thin wires without breaking. (See also **malleability.**)

electrical conductivity   the ability of a substance, such as a metal, or a solution to conduct an electrical current.

electrolyte   a substance that, when dissolved in water, dissociates into ions sufficiently to conduct an electrical current.

electron   a subatomic particle found in all neutral atoms; possesses the negative charges in atoms.

electron affinity   the energy released when a neutral atom gains an extra electron to form a negative ion.

electronegativity   the relative tendency of the atoms of an element to attract the electrons between two atoms in a chemical bond.

electronic configuration   a description of the arrangement of the electrons in an atom or ion, showing the numbers of electrons occupying each subshell.

electron neutrino   the variety of neutrino associated with electron emission in beta decay.

electrostatic   the type of interaction that exists between electrically charged particles; electrostatic forces attract particles together if the particles have charges of opposite sign, while the forces cause particles that have charges of like sign to repel each other.

element   a pure chemical substance that contains only one kind of atom.

emanation   the emission of radiation.

emission spectrum   the pattern of bright lines, as observed through a spectroscope, that results from atomic electrons transitioning to lower energy levels.

enrichment   an increase in the percentage of one isotope of an element compared with the percentage that occurs naturally.

excimer   an excited state of a diatomic molecule that has no ground state.

fallout   radioactive particles deposited from the atmosphere from either a nuclear explosion or a nuclear accident.

family   (See **group.**)

fission   (See **nuclear fission.**)

fluorescence   the spontaneous emission of light from atoms or molecules when electrons make transitions from states of higher energy to states of lower energy.

fusion   (See **nuclear fusion.**)

gamma decay   a mode of radioactive decay in which a very-high-energy photon of electromagnetic radiation—a gamma ray—is emitted; the daughter isotope has the same atomic number and mass number as the parent isotope, but lower energy.

gas centrifuge   a centrifuge in which gas flows continuously during operation.

gaseous diffusion   the process by which gases mix through random movement of the molecules in the gases.

Geneva Protocol   an international agreement consolidated after the end of WWI that prohibited the use of gas and bacteriological weapons in warfare.

global warming   the increase in the temperature of the oceans and lower atmosphere due to heating caused by absorption of infrared radiation emitted by Earth.

goiter   an enlargement of the thyroid gland, usually apparent as a swelling of the neck.

greenhouse effect   the warming effect produced in Earth's atmosphere whereby greenhouse gases in the atmosphere preferentially absorb and reemit infrared radiation.

greenhouse gas   an atmospheric gas that absorbs infrared radiation emitted by Earth and that contributes to warming of the atmosphere.

group   the elements that are located in the same column of the periodic table; also called a *family*, elements in the same column have similar chemical and physical properties.

half-life   the time required for half of the original nuclei in a sample to decay; during each half-life, half of the nuclei that were present at the beginning of that period will decay.

halide ion   a simple ion with a charge of -1 formed by adding an electron to a neutral halogen atom.

halogen   the elements in column VIIB of the periodic table; all of them share a common set of seven valence electrons in an nth energy level such that their outermost electronic configuration is ns^2np^5.

halogen halide   a chemical compound of the general form H-X, where "X" is a halogen atom: F, Cl, Br, or I.

halogenating agent   any chemical compound or ion that is capable of transferring halogen atoms to another compound or ion.

hepatitis   injury to the liver due to inflammation of its cells.

Hertzsprung-Russell diagram   used in astrophysics, a graph that plots luminosity versus surface temperature of a star.

hydrogen bond   a "bridge," or strong electrostatic attraction, between two molecules in which a hydrogen atom is shared by the two molecules; the hydrogen atom must be bonded to an atom of nitrogen, oxygen, or fluorine.

hydrogen halide   an acid containing a hydrogen atom and a halogen atom; HF is a weak acid, while the other acids—HCl, HBr, and HI— are strong acids.

hypothyroidism   the condition in which the thyroid gland produces insufficient quantities of the thyroid hormone.

inert an element that has little or no tendency to form chemical bonds.

inert gases an alternate name for the **noble gases.**

infalling materials falling into the atmosphere of a planet or star as a result of gravitational attraction.

interhalogen compound a binary chemical compound consisting of two different halogens.

ion an atom or group of atoms that have a net electrical charge.

ionic bond a strong electrostatic attraction between a positive ion and a negative ion that holds the two ions together.

isomeric state one of many forms of an element with the same mass and atomic number but different radioactive states.

isotope a form of an element characterized by a specific mass number; the different isotopes of an element have the same number of protons but different numbers of neutrons, hence different mass numbers.

lanthanide the elements ranging from cerium (atomic number 58) to lutetium (number 71); they all have two outer electrons in the 6s subshell plus increasingly more electrons in the 4f subshell.

laser an acronym for *l*ight *a*mplification of *s*timulated *e*mission of *r*adiation; a light amplifier used to produce ultraviolet, visible, or infrared light of a single frequency in a very narrow beam.

light helium the isotope of helium containing only one neutron in the nucleus.

luminescence the emission of light by an object for any reason besides an increase in temperature.

luminosity the brightness of a star defined as the total energy radiated per unit time.

luster the shininess associated with the surfaces of most metals.

main group element an element in one of the first two columns or one of the right-hand six columns of the periodic table; distinguished from transition metals, which are located in the middle of the table, and from rare earths, which are located in the lower two rows shown apart from the rest of the table.

main sequence   the area of the Hertzsprung-Russell diagram where most stars tend to be located during their early evolution.

malaria   an infectious disease transmitted by a parasite in which symptoms include chills, fever, sweating, nausea, and vomiting.

malleability   the ability of a substance such as a metal to change shape without breaking; metals that are malleable can be hammered into thin sheets. (See also **ductility.**)

marine acid   an obsolete name for hydrochloric acid.

mass   a measure of an object's resistance to acceleration; determined by the sum of the elementary particles composing the object.

mass number   the sum of the number of protons and neutrons in the nucleus of an atom. (See **isotope.**)

meridian   a line of longitude that runs from the north or south pole to the equator.

metal   any of the elements characterized by being good conductors of electricity and heat in the solid state; approximately 75 percent of the elements are metals.

metal halide   a chemical compound of the general form X-H, where "X" is a metal atom such as Li or Na.

metallicity   the extent to which a star contains elements heavier than helium.

metalloid (also called *semi-metal*)   any of the elements intermediate in properties between the metals and nonmetals; the elements in the periodic table located between metals and nonmetals.

methyl   in a molecule, a subgroup of carbon and hydrogen atoms with the general formula $-CH_3$.

monatomic   containing a single atom; the noble gases exist as monatomic species, and ions such as Na^+, K^+, and Ag^+ are said to be monatomic because they contain only one atom.

mottling   staining or discoloration of teeth or other tissues.

muriatic acid   a commercial name for hydrochloric acid.

muride   an obsolete name for bromine.

neutron   the electrically neutral particle found in the nuclei of atoms.

niton   an obsolete name for radon.

noble gas   any of the elements located in the last column of the periodic table—usually labeled column VIII or 18, or possibly column 0. Also called inert gas.

nonmetal   the elements on the far right-hand side of the periodic table that are characterized by little or no electrical or thermal conductivity, a dull appearance, and brittleness.

nonpolar solvent   a liquid that consists of nonpolar molecules; usually an oil or other hydrocarbon.

nuclear fission   the process in which certain isotopes of relatively heavy atoms such as uranium or plutonium break apart into fragments of comparable size; accompanied by the release of large amounts of energy.

nuclear fusion   the process in which certain isotopes of relatively light atoms such as hydrogen or helium can combine to form heavier isotopes; accompanied by the release of large amounts of energy.

nuclear medicine   the use of radioactive substances to image the body or to treat disease.

nucleon   a particle found in the nucleus of atoms; a proton or a neutron.

nucleosynthesis   the process of building up atomic nuclei from protons and neutrons or from smaller nuclei.

nucleus   the small, central core of an atom.

octet rule   a rule in chemistry associated with the tendency of atoms to gain or lose sufficient electrons so that their outer shell has eight electrons.

outgas   the emission of gas.

oxidation   an increase in an atom's oxidation state; accomplished by a loss of electrons or an increase in the number of chemical bonds to atoms of other elements. (See **oxidation state.**)

oxidation-reduction reaction   a chemical reaction in which one element is oxidized and another element is reduced.

oxidation state   a description of the number of atoms of other elements to which an atom is bonded. A neutral atom or neutral group of atoms of a single element are defined to be in the zero oxidation state. Otherwise, in compounds, an atom is defined as being in a positive or negative oxidation state, depending upon whether the atom is bonded to elements that, respectively, are more or less electronegative than that atom is.

oxidizing agent   a chemical reagent that causes an element in another reagent to be oxidized.

oxyanion   a negative ion that contains one or more oxygen atoms plus one or more atoms of at least one other element.

period   any of the rows of the periodic table; rows are referred to as periods because of the periodic, or repetitive, trends in the properties of the elements.

periodic table   an arrangement of the chemical elements into rows and columns such that the elements are in order of increasing atomic number, and elements located in the same column have similar chemical and physical properties.

persistent   referring to usually harmful substances that last for a very long time in the environment without breaking down into harmless substances.

phlogiston theory   an obsolete theory according to which flammable substances were thought to contain a substance called *phlogiston* that they released into the atmosphere during combustion.

photon   the name for the particle nature of light.

physical change   any transformation that results in changes in a substance's physical state, color, temperature, dimensions, or other physical properties; the chemical identity of the substance remains unchanged in the process.

physical state   the condition of a chemical substance being either a solid, liquid, or gas.

plasma   a state of matter where charged particles can move about freely.

polyatomic   a molecule or ion that contains two or more atoms.

polychlorinated biphenyl   a member of a class of compounds that contains two benzene rings linked together by a single bond and in which two or more hydrogen atoms have been replaced by chlorine atoms.

population inversion   the condition in a gas of atoms where more atoms are in an excited state preignition.

product   the compounds that are formed as the result of a chemical reaction.

proton   the positively charged subatomic particle found in the nuclei of atoms.

proton-proton chain   the chain reaction in stars like the Sun whereby hydrogen nuclei fuse to make helium nuclei.

quantum   a unit of discrete energy on the scale of single atoms, molecules, or photons of light.

radioactive decay   the disintegration of an atomic nucleus accompanied by the emission of a subatomic particle or gamma ray.

radium emanation   the radiation given off by radium.

rare earth element   the metallic elements found in the two bottom rows of the periodic table; the chemistry of their ions is determined by electronic configurations with partially filled f subshells. (See **lanthanide** and **actinide.**)

reactant   the chemical species present at the beginning of a chemical reaction that rearrange atoms to form new species.

red giant   the explosive evolutionary stage of a star that has fused all the hydrogen at its core into helium, which collapses under its own weight while hydrogen burning continues in the outer shell. The energy of the collapse generates radiation pressure that makes the outer shell expand explosively while the core continues to contract. The expanding gas in the outer shell filters out all but the star's red wavelengths.

reducing agent   a chemical reagent that causes an element in another reagent to be reduced to a lower oxidation state.

reduction   a decrease in an atom's oxidation state; accomplished by a gain of electrons or a decrease in the number of chemical bonds to atoms of other elements. (See **oxidation state.**)

salt lick   a naturally occurring salt deposit that provides animals with a source of salt.

sea salt   a soluble salt found in sea water; the salt found in highest quantity is sodium chloride.

semimetal   (See **metalloid**.)

shell   all of the orbitals that have the same value of the principal energy level, n.

specific heat   the heat per unit mass needed to raise the temperature of a substance by 1 C°.

spontaneous fission   the fission of a nucleus without the event having been initiated by human activity.

strong nuclear force   the force that binds together the particles inside an atom's nucleus.

subatomic particle   actual particles that are smaller than atoms (atoms are a composite), but not limited to atomic constituents. Neutrinos, for example, are subatomic particles.

sublimation   the change of physical state in which a substance goes directly from the solid to the gas without passing through a liquid state.

sublimation point temperature   the temperature of a pure substance at which the solid and gaseous phases become indistinguishable. Only carbon and arsenic are capable of sublimation.

sublime   (See **sublimation**.)

subshell   all of the orbitals of a principal shell that lie at the same energy level.

supernova   a colossal explosive event ending the evolution of a high-mass star and ejecting its matter into interstellar space.

synchrotron radiation   photons produced by the acceleration of charged particles.

thermal conductivity   a measure of the ability of a substance to conduct heat.

thyroid   a gland at the base of the neck that secretes hormones which control metabolic activities in the body.

tomogram, tomography   a medical term describing a two-dimensional image of a slice through a three-dimensional organ, tissue, etc.

transition metal   any of the metallic elements found in the 10 middle columns of the periodic table to the right of the alkaline earth metals; the chemistry of their ions largely is determined by electronic configurations with partially filled d subshells.

transmutation the conversion by way of a nuclear reaction of one element into another element; in transmutation, the atomic number of the element must change.

transuranium element any element in the periodic table with an atomic number greater than 92.

triad any group of three elements that exhibit very similar chemical and physical properties.

typhoid a bacteria-caused disease with such symptoms as high fever, sweating, intestinal disease, and diarrhea.

ultraviolet "beyond the violet"; the region of the electromagnetic spectrum that begins where violet light leaves off and is higher in energy and frequency than violet light.

vertebrate animals that have backbones and/or spinal columns.

volatile description of a liquid that evaporates readily at room temperature.

wavefront any collection of points on a wave having the same phase.

white dwarf the central, cool object that remains after an average star undergoes its red giant stage expansion that ejects most of the star's mass. A white dwarf is typically identified by its gravitational effects, as fusion processes can no longer occur.

Further Resources

The following is a list of sources that offer readings related to individual halogen and noble gas elements.

FLUORINE

Books and Articles

Banks, R. E. *Fluorine Chemistry at the Millennium.* Oxford, U.K.: Elsevier Science Ltd., 2000. A collection of contributions by leading researchers covering a breadth of understanding about fluorine chemistry.

Fagin, Dan. "Second Thoughts on Fluoride." *Scientific American* 298, no. 1 (January 2008): 74–81. This highly readable article gives an overall view of the fluoride debate.

Shriver, D. F., and P. W. Atkins. *Inorganic Chemistry.* 4th ed. New York: W.H. Freeman and Company, 2006. This book contains additional elementary information about the chemistry of fluorine and the other halogens.

Woosley, S. E., and W. C. Haxton. "Supernova Neutrinos, Neutral Currents and the Origin of Fluorine." *Nature* 334 (July 1988): 45–47. This article discusses astrophysical fluorine production in supernova explosions.

Internet Resources

National Institute for Dental and Craniofacial Research Web site. Available online. URL: http://www.nidcr.nih.gov. Accessed on November 8, 2007. This Web site contains considerable information about dental health and the benefits of fluoride in the prevention of tooth decay.

CHLORINE

Books and Articles

Carson, Rachel. *Silent Spring.* New York: Houghton-Mifflin Company, 1962. This is a classic environmental treatise against the use of pesticides.

Thornton, Joe. *Pandora's Poison: Chlorine, Health, and a New Environmental Strategy.* Cambridge, Mass.: The MIT Press, 2000. This book discusses the problem of organochlorines accumulating in the environment and proposes solutions.

Internet Resources

The United Nations Department for Disarmament Affairs. Available online. URL: http://www.un.org/Depts/dda/WMD/cwc/. Accessed on February 20, 2009. This Web site discusses the Chemical Weapons Convention that prohibits all development, production, acquisition, stockpiling, transfer, and use of chemical weapons.

BROMINE

Books and Articles

Graedel, Thomas E., and Paul J. Crutzen. *Atmosphere, Climate, and Change.* New York: Scientific American Library, 1995. Paul Crutzen was a co-recipient of the 1996 Noble Prize in chemistry for his work in atmospheric chemistry. In this book, the authors discuss significant global atmospheric problems such as global warming and the threats to stratospheric ozone.

Ornes, Stephen. "The Hole Story." *Discover* (February 2007): 60–61. This article discusses possible reasons for the enlargement of the ozone hole above the South Pole.

Internet Resources

Leutwyler, Kristin. Available online. "PET Scans Show Metabolic Changes that Herald Future Memory Losses." *Scientific American,* September 11, 2001. URL: http://www.scientificamerican.com/article.cfm?id=pet-scans-show-metabolic&modsrc =related_links. This is an article about how PET scans may help identify people at risk for Alzheimer's disease.

IODINE AND ASTATINE

Books and Articles

Brownstein, David. *Iodine: Why You Need It, Why You Can't Live without It.* West Bloomfield, Mich.: Medical Alternatives Press, 2006.

A short, nontechnical discourse on the importance of iodine in human cells.

Dajer, Tony. "Vital Signs: A Young Woman's On-Again, Off-Again Pain Has an Unexpected Origin." *Discover* (February 2007): 30–31. This article describes the sometimes tricky diagnosis of hyperthyroidism.

Smith, J. Michael, et al. "United States Population Dose Estimates for Iodine 131 in the Thyroid after the Chinese Atmospheric Nuclear Weapons Tests." *Science 7,* 200, no. 4,337 (April 1978): 44–46. This article elucidates how nuclear weapons tests in one country can have a global effect.

Swain, Patricia A. "Bernard Courtois (1777–1838), Famed for Discovering Iodine (1811), and His Life in Paris from 1798." *Bull. Hist. Chem.* 30, no. 2 (2005): 103–111. This article describes the discovery of iodine.

Van Nostrand, Douglas, et al. "Side Effects of 'Rational Dose' Iodine 131 Therapy for Metastatic Well-Differentiated Thyroid Carcinoma." *Journal of Nuclear Medicine* 27 (1986): 1,519–1,527. This journal article gives data on the rate of side effects in patients undergoing iodine 131 therapy for thyroid cancer. It is accessible to the lay reader.

HELIUM
Books and Articles

Botting, Douglas. *The Giant Airships.* Alexandria, Va.: Time-Life Books, 1981. Part of a multivolume series on the history of flight, this volume highlights the history of dirigibles.

Eskridge, Brent, and Dwight Neuenschwander. "A Pedagogical Model of Primordial Helium Synthesis." *American Journal of Physics* 64, no. 12 (December 1996): 1,517–1,524. The calculation of the primordial hydrogen and helium abundances in the big-bang cosmology is presented in an oversimplified model accessible to university physics students who have had no physics beyond an elementary modern physics course.

Kluger, Jeffrey. "Boy oh Buoyant." *Discover,* December 1995, 50. This article discusses the logistics of managing helium when flying a blimp.

Internet Resources

Los Alamos National Laboratory. Available online. URL: http://www.lanl.gov/news/index.php/fuseaction/home.story/story_id/10503. Accessed on March 13, 2009. This article describes how helium slows the solar wind.

Williams, Mark. "Mining the Moon." *Technology Review* (August 2007). Available online. URL: www.technologyreview.com/Energy/19296. Accessed on October 28, 2009. This article discusses whether future fusion reactors could use helium 3 gathered from the Moon.

NEON

Books and Articles

Miller, Samuel, and Wayne Strattman. *Neon Techniques*. Cincinnati, Ohio: ST Media Group, 2003. This is a highly readable book on the history and current usage of the luminous tube.

Schiff, Eric. "How Neon Lights Work." *Scientific American,* April 17, 2006. This article explains the physics of lighted neon.

Townes, Charles. *How the Laser Happened: Adventures of a Scientist.* New York: Oxford Press, 1999. This book provides a comprehensive history of laser science.

ARGON

Books and Articles

Cook, Gerhard A. *History, Occurrence, and Properties.* Argon, Helium and the Rare Gases: The Elements of the Helium Group, vol. 1. New York: Interscience Publishers, 1961. This book details what was known about the helium group in the mid-20th century.

Khriachtchev, Leonid, et al. "A Stable Argon Compound." *Nature,* 24 August 2000, 874–876. This article describes how to synthesize argon fluorohydride.

Internet Resources

The American Cancer Society. Available online. URL: http://www.cancer.org/docroot/NWS/content/NWS_3_1x_Freezing_Tumors_Compares_Favor ably_to_Traditional_Treatments.asp. Accessed

March 14, 2009. This Web site gives information on freezing tumors as an option for men with prostate cancer.

KRYPTON AND XENON
Books and Articles

Cullen, Stuart C., and Erwin G. Gross. "The Anesthetic Properties of Xenon in Animals and Human Beings, with Additional Observations on Krypton." *Science* (May 1951): 18.

Owen, T., K. Biemann, D. R. Rushneck, J. E. Biller, D. W. Howarth, and A. L. Lafleur. "The Atmosphere of Mars: Detection of Krypton and Xenon." *Science* (December 1976): 194.

RADON
Books and Articles

Cole, Leonard. *Element of Risk: The Politics of Radon.* Oxford, England: Oxford University Press, 1994. The book discusses political debates, media issues, and the ways to decide about radon and lung cancer in the home.

Farley, K. A., and A. Neroda. "Noble Gases in the Earth's Mantle." *Annual Review of Earth and Planetary Science* 26 (May 1998): 189–218. This chapter describes the sources of the noble gases in Earth's mantle and describes some of the current controversy about the origins of observed isotopic ratios.

Makofske, William J., and Michael R. Edelstein. *Radon and the Environment.* Park Ridge, N.J.: Noyes Publications, 1988. This is an overview of radon hazards, health, and technology—a bit dated, but rigorous.

General Resources

The following is a list of sources that discuss general information on the periodic table of the elements.

Books and Articles

Ball, Philip. *The Elements: A Very Short Introduction.* Oxford: Oxford University Press, 2002. This book contains useful information about the elements in general.

Chemical and Engineering News 86, no. 27 (2 July 2008). A production index is published annually showing the quantities of various chemicals that are manufactured in the United States and other countries.

Considine, Douglas M., ed. *Van Nostrand's Encyclopedia of Chemistry,* 5th ed. New York: Wiley, 2005. In addition to its coverage of traditional topics in chemistry, the encyclopedia has articles on nanotechnology, fuel cell technology, green chemistry, forensic chemistry, materials chemistry, and other areas of chemistry important to science and technology.

Cotton, F. Albert, Geoffrey Wilkinson, and Paul L. Gaus. *Basic Inorganic Chemistry,* 3rd ed. New York: Wiley, 1995. Written for a beginning course in inorganic chemistry, this book presents information about individual elements.

Cox, P. A. *The Elements on Earth: Inorganic Chemistry in the Environment.* Oxford: Oxford University Press, 1995. There are two parts to this book. The first part describes Earth and its geology and how elements and compounds are found in the environment. It also describes how elements are extracted from the environment. The second part describes the sources and properties of the individual elements.

Daintith, John, ed. *The Facts On File Dictionary of Chemistry,* 4th ed. New York: Facts On File, 2005. Extensive collection of definitions on the many technical terms used by chemists.

Emsley, John. *Nature's Building Blocks: An A–Z Guide to the Elements.* Oxford: Oxford University Press, 2001. Proceeding through the periodic table in alphabetical order of the elements, Emsley describes each element's important properties, biological and medical roles, and importance in history and the economy.

———. *The Elements.* Oxford: Oxford University Press, 1989. In this book, Emsley provides a quick reference guide to the chemical, physical, nuclear, and electron shell properties of each of the elements.

Greenberg, Arthur. *Chemistry: Decade by Decade.* New York: Facts On File, 2007. An excellent book that highlights by decade the important events that occurred in chemistry during the 20th century.

Greenwood, N. N., and A. Earnshaw. *Chemistry of the Elements.* Oxford, U.K.: Pergamon Press, 1984. This book is a comprehensive treatment of the chemistry of the elements.

Hall, Nina, ed. *The New Chemistry.* Cambridge: Cambridge University Press, 2000. This book contains chapters devoted to the properties of metals and electrochemical energy conversion.

Hampel, Clifford A., ed. *The Encyclopedia of the Chemical Elements.* New York: Reinhold Book Corp., 1968. In addition to articles about individual elements, this book also has articles about general topics in chemistry. Numerous authors contributed to this book, all of whom were experts in their respective fields.

Heiserman, David L. *Exploring Chemical Elements and Their Compounds.* Blue Ridge Summit, Pa.: Tab Books, 1992. This book is described by its author as "a guided tour of the periodic table for ages 12 and up," and is written at a level that is very readable for precollege students.

Henderson, William. *Main Group Chemistry.* Cambridge, U.K.: The Royal Society of Chemistry, 2002. This book is a summary of inorganic chemistry in which the elements are grouped by families.

Jolly, William L. *The Chemistry of the Non-Metals.* Englewood Cliffs, N.J.: Prentice Hall, 1966. This book is an introduction to the chemistry of the nonmetals, including the elements covered in this book.

King, R. Bruce. *Inorganic Chemistry of Main Group Elements.* New York: Wiley-VCH, 1995. This book describes the chemistry of the elements in the s and p blocks.

Krebs, Robert E. *The History and Use of Our Earth's Chemical Elements: A Reference Guide,* 2nd ed. Westport, Conn.: Greenwood Press, 2006. Following brief introductions to the history of chemistry and atomic structure, Krebs proceeds to discuss the chemical and physical properties of the elements group (column) by group. In addition, he describes the history of each element and current uses.

Lide, David R. ed. *CRC Handbook of Chemistry and Physics,* 89th ed. Boca Raton, Fla.: CRC Press, 2008. The *CRC Handbook* has been the most authoritative, up-to-date source of scientific data for almost nine decades.

Mendeleev, Dmitri Ivanovich. *Mendeleev on the Periodic Law: Selected Writings, 1869–1905.* Mineola, N.Y.: Dover, 2005. This English translation of 13 of Mendeleev's historic articles is the first easily accessible source of his major writings.

Minkle, J. R. "Element 118 Discovered Again—For the First Time." *Scientific American,* 17 October 2006. This article describes how scientists in California and Russia fabricated element 118.

Norman, Nicolas C. *Periodicity and the p-Block Elements.* Oxford: Oxford University Press, 1994. This book describes group properties of post-transition metals, metalloids, and nonmetals.

Parker, Sybil P., ed. *McGraw-Hill Encyclopedia of Chemistry,* 2nd ed. New York: McGraw Hill, 1993. This book presents a comprehensive treatment of the chemical elements and related topics in chemistry, including expert-authored coverage of analytical chemistry, biochemistry, inorganic chemistry, physical chemistry, and polymer chemistry.

Rouvray, Dennis H., and R. Bruce King, ed. *The Periodic Table: Into the 21st Century.* Baldock, Hertfordshire, U.K.: Research Studies Press Ltd., 2004. A presentation of what is happening currently in the world of chemistry.

Stwertka, Albert. *A Guide to the Elements,* 2nd ed. New York: Oxford University Press, 2002. This book explains some of the basic concepts of chemistry and traces the history and development of the periodic table of the elements in clear, nontechnical language.

Van Nostrand's Encyclopedia of Chemistry, 5th ed., Glenn D. Considine, ed. Hoboken, N.J.: Wiley, 2005. A valuable reference work that is concise in its one-volume approach to coverage of chemical topics.

Winter, Mark J., and John E. Andrew. *Foundations of Inorganic Chemistry.* Oxford: Oxford University Press, 2000. This book presents an elementary introduction to atomic structure, the periodic table, chemical bonding, oxidation and reduction, and the chemistry of the elements in the s, p, and d blocks; in addition, there is a separate chapter devoted just to the chemical and physical properties of hydrogen.

Internet Resources

American Chemical Society. Available online. URL: www.chemistry. org. Accessed December 19, 2008. Many educational resources are available online.

Center for Science and Engineering Education, Lawrence Berkeley Laboratory, Berkeley, California. Available online. URL: www.lbl. gov/Education. Accessed October 28, 2009. Contains educational resources in biology, chemistry, physics, and astronomy.

Chemical Education Digital Library. Available online. URL: www. chemeddl.org/index.html. Accessed December 19, 2008. Digital content intended for chemical science education.

Chemical Elements. Available online. URL: www.chemistryexplained. com/elements. Accessed December 19, 2008. Information about each of the chemical elements.

Chemical Elements.com. Available online. URL: www.chemical elements.com. Accessed December 19, 2008. A private site that originated with a school science fair project.

Chemicool, created by David D. Hsu of the Massachusetts Institute of Technology. Available online. URL: www.chemicool.com. Accessed December 19, 2008. Information about the periodic table and the chemical elements.

Department of Chemistry, University of Nottingham, United Kingdom. Available online. URL: www.periodicvideos.com. Accessed December 19, 2008. Short videos on all of the elements can be viewed. The videos can also be accessed through YouTube®.

Journal of Chemical Education, Division of Chemical Education, American Chemical Society. Available online. URL: jchemed.chem.wisc.edu/HS/index.html. Accessed December 19, 2008. The Web site for the premier online journal in chemical education.

Lenntech Water Treatment & Air Purification. Available online. URL: www.lenntech.com/Periodic-chart.htm. Accessed December 19, 2008. Contains an interactive, printable version of the periodic table.

Los Alamos National Laboratory, Chemistry Division, Los Alamos, New Mexico. Available online. URL: periodic.lanl.gov/default.htm. Accessed December 19, 2008. A resource on the periodic table for elementary, middle school, and high school students.

Mineral Information Institute. Available online. URL: www.mii.org. Accessed December 19, 2008. A large amount of information for teachers and students about rocks and minerals and the mining industry.

National Nuclear Data Center, Brookhaven National Laboratory, Upton, N.Y. Available online. URL: http://www.nndc.bnl.gov/content/HistoryOfElements.html. Accessed December 19, 2008. A worldwide resource for nuclear data.

The New York Times Company, About.com. "Chemistry." Available online. URL: chemistry.about.com/od/chemistryfaqs/f/element.htm. Accessed December 19, 2008. Information about the periodic table, the elements, and chemistry in general.

The Periodic Table of Comic Books, Department of Chemistry, University of Kentucky. Available online. URL: www.uky.edu/Projects/Chemcomics. Accessed December 19, 2008. A fun, interactive version of the periodic table.

The Royal Society of Chemistry. URL: http://www.rsc.org/chemsoc/. Accessed January 17, 2009. This site contains information about many aspects of the periodic table of the elements.

Schmidel & Wojcik: Web Weavers. Available online. URL: quizhub. com/quiz/f-elements.cfm. Accessed December 19, 2008. A K–12 interactive learning center that features educational quiz games for English language arts, mathematics, geography, history, earth science, biology, chemistry, and physics.

United States Geological Survey. Information available online. URL: minerals.usgs.gov. Accessed December 19, 2008. The official Web site of the Mineral Resources Program.

Web Elements, The University of Sheffield, United Kingdom. Available online. URL: www.webelements.com/index.html. Accessed December 19, 2008. A vast amount of information about the chemical elements.

Wolfram Science. Available online. URL: demonstrations.wolfram. com/PropertiesOfChemicalElements. Accessed December 19, 2008. Information about the chemical elements from the Wolfram Demonstration Project.

Periodicals

Discover
Published by Buena Vista Magazines
114 Fifth Avenue
New York, NY 10011
Telephone: (212) 633-4400
www.discover.com
A popular monthly magazine containing easy-to-understand articles on a variety of scientific topics.

Nature
The Macmillan Building
4 Crinan Street
London N1 9XW
Telephone: +44 (0)20 7833 4000
www.nature.com/nature
A prestigious primary source of scientific literature.

Science
Published by the American Association for the Advancement of
  Science
1200 New York Avenue NW
Washington, DC 20005
Telephone: (202) 326-6417
www.sciencemag.org
One of the most highly regarded primary sources for scientific
  literature.

Scientific American
415 Madison Avenue
New York, NY 10017
Telephone: (212) 754-0550
www.sciam.com
A popular monthly magazine that publishes articles on a broad range
  of subjects and current issues in science and technology.

Index